IONIZATION IN DIESEL COMBUSTION: MECHANISM, NEW INSTRUMENTATION AND ENGINE APPLICATIONS

by

FADI A. ESTEFANOUS

DISSERTATION

Submitted to the Graduate School

of Wayne State University

Detroit, Michigan

in partial fulfillment of the requirements

for the degree of

DOCTOR OF PHILOSOPHY

2011

MAJOR: MECHANICAL ENGINEERING

Approved by:

_____

Advisor   Date

_____

_____

_____

UMI Number: 3466645

Dissertation Publishing

ProQuest LLC
789 East Eisenhower Parkway
P.O. Box 1346
Ann Arbor, MI 48106-1346

ACKNOWLEDGMENTS

I would like to express my deep and sincere gratitude to my adviser Dr. Naeim A. Henein for trusting me and giving me an opportunity to be a part of his research team at Wayne State University. He always guided and supported me for many years. He taught me hard work, positive thinking, problem solving, dedication and most importantly, how to convert failure into success. This, I will cherish all my life. Under his supervision, I enjoyed my time at Wayne State University, both as a student and as a graduate research assistant. I could not have made it without him.

I thank Dr. Marcis Jansons for helping me in conducting experiments on the optically accessible engine and for his positive feedbacks and support. I also thank Tamer Badawy for assisting me in running tests on the John Deere heavy duty diesel engine. I acknowledge help from Lidia Nedeltcheva for providing technical support, Eugene, Marvin and Dave from the machine shop for machining the components of ion sensor with high precision.

I would like to acknowledge the tremendous effort and invaluable support of my wife Marie who helped me throughout the course of my Ph.D. She has always been there for me, inspiring and driving me towards a better future. Without her remarkable sacrifices, it would have been certainly much harder to finish my dissertation. I also owe my love to my daughter Eliana. Finally, I am forever indebted to my parents and my sister Catherine for their encouragement when it was most required.

TABLE OF CONTENTS

CHAPTER 1   INTRODUCTION

Diesel engines are known for their superior fuel economy and high power density. However they emit undesirable high levels of nitrogen oxide (NO_x) and black particulate smoke (Soot) [1- 3]. These engines are required to meet stringent emissions standards, while still improving performance and reducing fuel consumption.   To meet these demands, close loop engine control strategies have been developed based on signals from several sensors that indirectly indicate the in-cylinder combustion process.

The effectiveness of the control strategies in fine tuning the engine parameters would significantly improve if an in-cylinder direct combustion sensor is developed [4]. Of the methods known for detecting combustion conditions during engine operation, the ion current is considered to be highly useful because it can be used an indicator chemical species resulting from the engine combustion [5- 9]. As such, an in-cylinder ionization sensor may be employed to sense various engine parameters under different engine operating conditions [10- 23]. For instance, the ion current can indicate some combustion parameters, based on the fact that positive and negative ions and electrons are generated during the combustion process [6, 24- 27].

In gasoline engines, for instance, spark plugs are used as ion current sensors [28-35]. In diesel engines, glow plugs are modified and used as ion current sensors [36- 39]. However, the use of the modified glow plug is limited to engines fitted with glow plugs. Other engines have to drill a hole in the cylinder head to accommodate a glow plug.

The ion current has been well investigated in spark ignited engines, where the combustible mixture is homogeneous. However, limited research has been carried out and published in the area of ionization in diesel engines [36, 40- 43] where the mixture is heterogeneous and the combustion process is complex.

In order to use the ion current signal in diesel engines control, this signal has to be better understood, simulated, and predicted using adequate mathematical models under different engine operating conditions. In this dissertation new experimental and analytical techniques are developed and implemented. In addition, a new diesel ion formation chemical kinetics mechanism is developed. This mechanism is then implemented in a three dimensional CFD code taking into consideration fuel injection, atomization, droplet and liquid film evaporation, and mixture heterogeneity. The model predictions are compared with experimental results obtained in a multi-cylinder diesel engine under different operating conditions.

CHAPTER 2   LITERATURE REVIEW

2.1 CHAPTER OVERVIEW

This chapter gives a comprehensive review literature review of the research on ionization in diesel combustion.   The review covers ionization in different types of flames, negative and positive ions, soot ionization, models developed and the reactions for thermal and chemi ionization.

2.2 INVESTIGATION ON POSITIVE IONS IN FLAMES

Van Tiggelen, Deckers and Jaegere [44] tried to identify the most abundant ions in some flames, and found that in acetylene and oxygen flames H_3O^+ is the most abundant ion representing 70 to 90 % of the total ions. $C_3H_3^+$ was found only in rich acetylene flames. NO^+ does not appear when nitrogen is not added to the flammable mixture. Very slight peaks were observed in order of decreasing importance as follows: HCO^+, $C_2H_3O^+$, CO^+, $C_2H_2^+$, C_2^+, CH^+, CH_2^+, CH_3^+ and OH^+. NO^+ is the most abundant ion, almost 90 % of the total ion concentration, in nitrogen oxide producing flames followed by H_3O^+ and NH_4^+, $C_3H_3^+$ only in rich mixtures and slight peaks caused by HCO^+, $C_2H_2^+$, NOH^+ and $C_2H_3O^+$. NO^+ and NH_4^+ are the major ions in $C_2H_2{:}NH_3{:}O_2$ flames.   Bertand and Tiggelen [45] investigated ions in ammonia flames and observed NO^+, NH_4^+ and N_3O^+ in $NH_3{:}O_2$ mixtures. By adding H_2 and N_2 to the flammable mixture they obtained NO^+ (80%) and H_3O^+ (20%). They tried a third mixture of $NH_3{:}N_2{:}CO{:}O_2$ and found only NO^+.

2.3 SOURCE OF NITRIC OXIDE IONS

Sugden [49] studied the source of NO^+ ions in flames containing hydrocarbon additives. He proposed the source of NO^+ formed in these flames based on measurements of the concentration profiles of NO^+, O_2^+, CHO^+, and H_3O^+.

CO flames showed two regions of NO^+ formation, a primary one in the reaction zone and a secondary one downstream of it. He found that NO^+ formation in the secondary region is linked to that of O_2^+ ion. NO addition results in the rapid replacement of O_2^+ by NO^+, and therefore the reaction $(O_2^+ + NO => NO^+ + O_2)$ was proposed. However, NO^+ formation in the primary region is shown to be consistent with charge exchange with the CHO^+ ion, and the reaction $(CHO^+ + NO => NO^+ + CHO)$ was suggested. Formation of NO^+ in hydrogen flames with NO addition is restricted only in the primary zone.

Sugden made a noticeable remark in this paper on hydrogen and CO flames. He found the value of $[NO^+]$ at its maximum greatly exceeds that of Saha equilibrium at the same temperatures used, 2400 K in this case. Also the rate of attainment of this value is likewise too fast for a process of direct thermal ionization of nitric oxide.

Goodings and Bohme [50] studied the formation of NO^+ in methane-air flames. They mentioned that the production mechanism for NO^+ is not straight forward. Despite the low ionization potential of NO, the amount of equilibrium thermal ionization calculated using Saha equation accounts for only about 1% of the NO^+ observed during the experiments.

Lin and Teare [51] studied air ionization in shock waves tube. They categorized the predominant electron production process in the following order, atom-atom ionizing collision, then atom-molecule collision, molecule-molecule collision. They also made a list of reactions for the oxygen-nitrogen system as follows

$$N + O + 2.8eV \rightleftharpoons NO^+ + e,$$
$$N + N + 5.8eV \rightleftharpoons N_2^+ + e,$$
$$N + O_2 + 6.5eV \rightleftharpoons NO_2^+ + e,$$
$$O + O + 6.9eV \rightleftharpoons O_2^+ + e,$$
$$O + NO + 7.9eV \rightleftharpoons NO_2^+ + e,$$
$$N + NO + 7.9eV \rightleftharpoons N_2O^+ + e,$$
$$X + NO + 9.3eV \rightleftharpoons X + NO^+ + e,$$
$$O + N_2 + 11.2eV \rightleftharpoons N_2O^+ + e,$$
$$N_2 + O_2 + 11.2eV \rightleftharpoons NO + NO^+ + e,$$
$$O + O_2 + 11.7eV \rightleftharpoons O_3^+ + e,$$
$$X + O_2 + 12.1eV \rightleftharpoons X + O_2^+ + e,$$
$$X + O + 13.6eV \rightleftharpoons X + O^+ + e,$$
$$X + N + 14.6eV \rightleftharpoons X + N^+ + e,$$
$$X + N_2 + 15.6eV \rightleftharpoons X + N_2^+ + e.$$

They stated that among the previous reactions, the N-O reaction appears to be the major contributor of NO^+ formation.

Tiggelen [45] proposed that chemi-ionization involving ground state oxygen atom and excited electronic state nitrogen lead to the formation of NO^+ in NH_3-oxygen flames. Hansen [52] studied NO ionization processes at fairly high temperatures, and found that the most important mechanism is the collision between N and O atoms. So he also favored the same reaction suggested by Lin and Teare.

Bulewicz and Padley [53] showed that even in hot cyanogens-oxygen flame, chemi but not thermal ionization is the main source of ions. Fialkov and Kalinich [54] found that NO^+ concentration decreases as fast as the concentration of oxygen decreases in propane-air and benzene-air flames. They found that when oxygen is absent in the surrounding gas, NO^+ is not formed. They also mentioned that the concentration of this ion with the equivalence ratio increases up to the appearance of the yellow soot luminosity. Their findings correlate well with the suggestion that ($N + O => NO^+ + e$) reaction is one of the major sources of NO^+ formation.

2.4 INVESTIGATION ON NEGATIVE IONS IN FLAMES

Negative ions have been reviewed as well through the literature in order to compare its concentration with that of positive ions and electrons. Goodings and Bohme [55] studied the negative ion chemistry in methane-oxygen flames and found that downstream in the fuel lean flame the concentration of negative ions and free electrons are equal. However, in fuel rich flames, downstream, negative ions disappear making positive ions concentration equal to the free electrons concentration. As a conclusion, the negative ion profile decrease sharply when the concentration of hydroxyl (OH), oxygen (O), and hydrogen (H) radicals are rising rapidly. These are natural reagents for the loss of negative ions in associative detachment processes.

Hayhurst and Kittelson [56] studied oxy-acetylene flames and came up with the same conclusions as Goodings. They stated that in every flame the total negative ion current maximizes near the upstream edge of the visible reaction zone and the positive and negative ion currents are equal. Downstream of this point, negative ions decay rapidly while the positive ion current continues to rise to a maximum two to three times that for negative ions.

Lin and Teare [51] studied the rate of ionization behind shock wave tubes in air and found that the concentration of O_2^- and O^- would always be negligible in comparison with the electron concentration when the gas temperature exceeds about 1500 K.

Brown and Eraslan [57] agreed to not include negative ions in their chemical kinetic model based on the fact that most of the negative charge consists of free electrons, especially downstream of the reaction zone.

Fialkov [58] performed ion investigations on flames and proclaimed that negatively charged species are represented by electrons and negative ions and that it is very difficult to determine concentrations shares of each one of them. It is obvious that even if, say, half of the electrons are attached to species, the flame electric properties could be interpreted as if only free electrons are in the flame. Therefore, the equation $n_+ = n_e$, where "n" is the number density of the species, is often adopted. He also stated that the absolute concentrations of negative ions were quite different, depending on the method of measuring. For example, Langmuir probes showed same concentration of positive and negative ions with electrons concentration only 2% of the positive concentration. However, mass spectrometer gave concentration of negative ions two order of magnitude less than positive ions.

2.5 MATHEMATICAL MODELS FOR IONS IN FLAMES

Ion-signal calculation using mathematical and chemical-kinetics modeling has been conducted over the past years in an attempt to better understand and predict the ion-current and thus use this signal as a method of engine control. In the next few lines, the difference between several ion-current models will be shown.

In 1988, Brown and Eraslan [57], from Iowa State University, simulated lean and close to stoichiometric acetylene flames using a chemical kinetic model. The set of reactions included the oxidation, pyrolysis, chemi-ionization, ion molecule reactions and charge recombination. Ions included in the model were H_3O^+, CHO^+, CH_3^+, CH_2O^+, CH_3O^+, CH_5O^+, C_2HO^+, $C_2H_3O^+$, $C_3H_3^+$, $C_5H_3^+$, $C_5H_5^+$, and $C_7H_5^+$. They found H_3O^+, $C_3H_3^+$ and $C_2H_3O^+$ to be the three principal ions in this flame, and stated that ions of very large mass, greater than 300 amu, can appear, especially as the flame is made richer, as the critical equivalence ratio for soot formation is approached. They also published another paper during the same year [59] which modeled the chemi-ionization reactions in acetylene fuel rich flame and concluded that the source of ions in fuel rich-flame was not clear.

2.6 MATHEMATICAL MODELS FOR IONS IN INTERNAL COMBUSTION ENGINES

In 1996, Saitzkoff and Reinmann [5] from Lund Institute of technology simulated the ion current in spark ignited engines. They used the spark plug as an ion sensor and noticed that the ion current has three distinct phases. First a large current that generates a spark and ignites the mixture, then a current peak early in the combustion that does not seem to be directly related to the pressure and finally a current that seems to be directly related to the pressure in the cylinder. Their purpose was to explain the processes responsible for the last phase. An analytical expression for the ion current as function of temperature was derived using Saha's equation as follows:

$$I \; = \; A \sqrt{\frac{\varPhi_s}{n_{tot}}} \; T^{\frac{1}{4}} \exp\left[-\frac{E_i}{2kT}\right] \qquad (1)$$

Where, ($\varPhi_s$) is the species mole fraction, (E_i) is the ionization energy of the same species, (n_{tot}) is the total number density, and the symbol (A) is a constant.

They found that a relatively minor species, NO, seems to be the major agent responsible for the conductivity of the hot gas in the spark gap. They proposed the mechanism of NO ionization to be only thermal, which means that it is only the high temperature that causes NO to ionize as follows ($NO \Rightarrow NO^+ + e$). In order for them to match the experimental data with the calculated based on NO thermal ionization, they assumed NO concentration of 10,000 ppm at the adiabatic flame temperature of 2800 K and maximum pressure of 5.7 MPa. The results are shown in FIG 2.1.

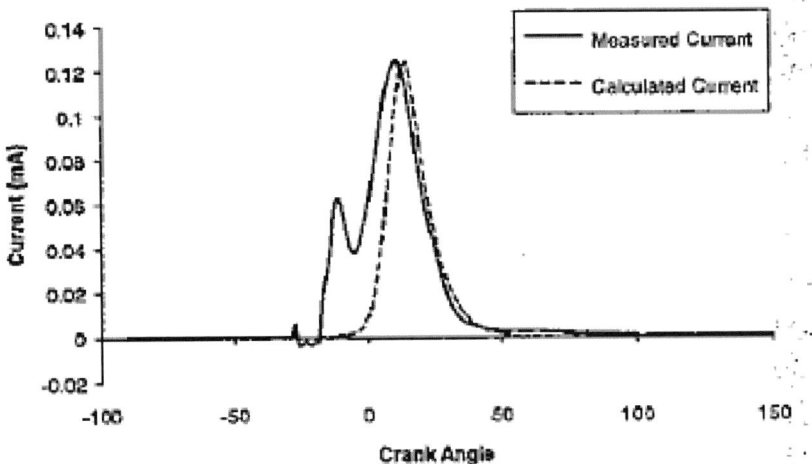

Figure 2.1   Comparison between the measured current and the calculated current [5].

Reinmann and Saitzkoff [24] measured the ion current and built a chemical kinetic model to predict the local air to fuel ratio in the vicinity of the spark plug in gasoline engines based on the chemiionization reactions which takes place within the first ion-current peak, known as the reaction zone peak. The chemical model used was able to estimate only the ion current in the reaction zone and was burning a mixture of iso-octane and   n-heptane, and was predicting a correlation between H_3O^+ and the air/fuel ratio (λ).

$$\left[H_3O^+ \right]_{max} - \frac{const}{\sqrt{\lambda}} \qquad (2)$$

The ions used within the model were CHO^+, H_3O^+, and $C_3H_3^+$. A comparison between the experimental data and the model calculation results for the first ion current peak is shown in FIG 2.2, and it is obvious that there is a discrepancy which increased by increasing the equivalence ratio.

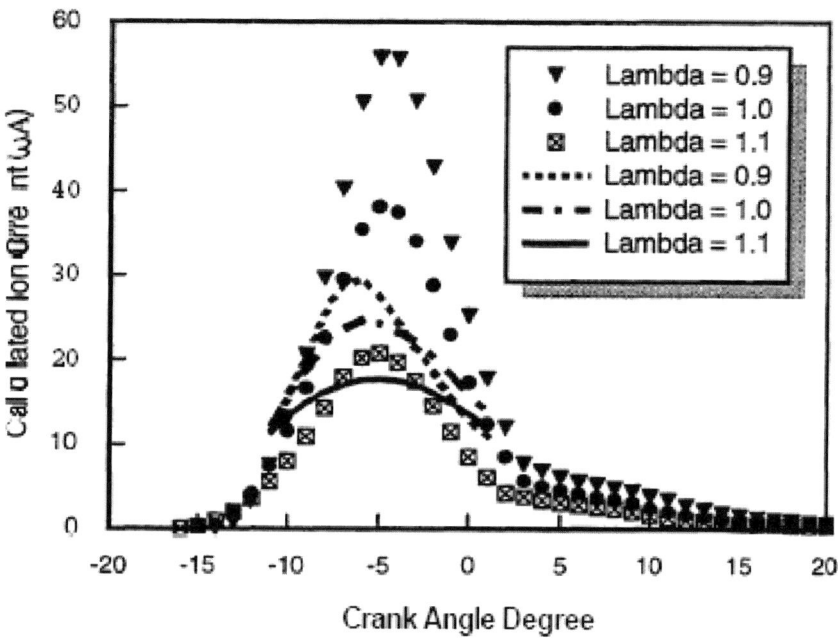

Figure 2.2   Ion currents calculated from the chemical kinetic model are indicated by the lines and the measured ion currents are represented by the symbols according to the legend [24].

In another paper [6] in 1997, the same group talked again about the effect of NO in the post flame zone concluding that thermal ionization is the governing ion formation process in this region. They built a simple zero-dimensional model with chemical kinetics based on 64 species and 268 reactions in order to come up with the neutral species concentration in the post flame zone. For the NO calculation in the model, the reactions in the extended Zeldovich mechanism were used. The NO concentration predicted from that model was 14,800 ppm and was used as an input to the thermal ionization model in order to obtain the ionization ratio and thus the ion current.

They also talked about the role of the negative ions and electrons in carrying the ion current and concluded, based on their calculations, that even if the major part of the electrons are attached and only a minority remain free, it is still the electrons that are responsible for the most of the current due to their much lower mass and therefore much higher drift velocity as shown in FIG 2.3.

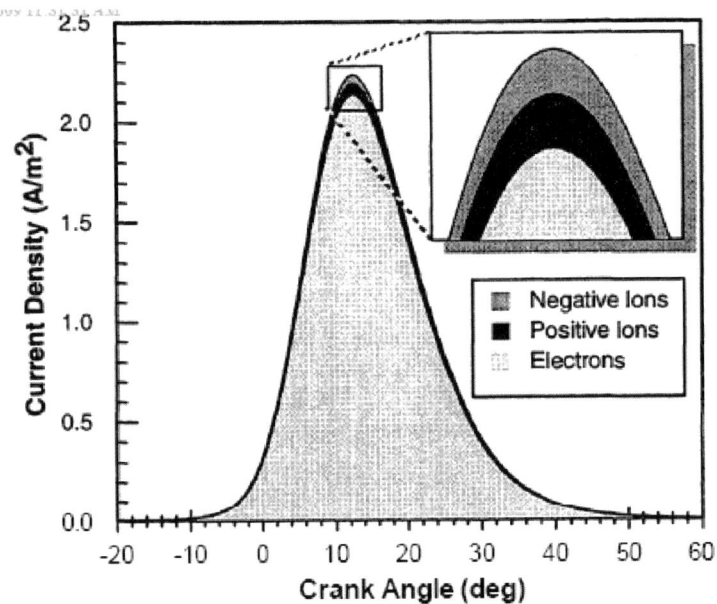

Figure 2.3   The calculated cumulative current contribution from electrons, positive and negative ions under stoichiometric condition [6].

In 1998, the same group published a paper [60] which discusses the different processes of ion formation in flames using different fuels in a spark ignited engine. They noticed two ion current peaks and as previously mentioned, the first peak is due to chemi-ionization and the second peak is due to the thermal ionization of NO. They stated that the thermal equilibrium analysis for the second peak, performed in previous studies, was based on quite high concentrations of NO. Most exhaust gas measurements indicated that the values used for the investigation seemed to be in the order of 3 to 10 times higher than the measured ones. They claimed that this difference arises from the fact that the local NO concentration around the spark plug may reach much higher values than the average concentration over the whole cylinder as the NO formation process is slow in comparison to other flame reactions.

Naoumov and Demin [61] simulated the ion current in spark ignited engines using a chemical kinetic model. The model of fuel and combustion included 65 species and 247 reactions, and they used for gasoline vapor fuel simulation a mixture of octane (C_8H_{18}) and methylcyclohexane ($C_6H_{11} - CH_3$) and toluene ($C_6H_5 - CH_3$). For chemi and thermal ionization they used a total of 6 reactions including CHO^+, H_3O^+, NO^+, and electrons. They stated the NO thermal ionization process as an irreversible chemical reaction with only a forward reaction rate as shown below.

chemi-ionozation
1. $CH + O \Leftrightarrow CHO^+ + e$
charge transfer reactions
2. $CHO^+ + H_2O \Leftrightarrow H_3O^+ + CO$
3. $CHO^+ + NO \Leftrightarrow NO^+ + HCO$
charge recombination
4. $H_3O^+ + e \Leftrightarrow 2H + OH$
5. $NO^+ + e \Leftrightarrow N + O$
ionization by collision + thermo- ionization
6. $NO + M \Rightarrow NO^+ + e + M$

Compared to the previous work done by Saitzkoff which only described the second peak, their model was able to capture both first and second peaks although the amplitudes does not completely match as shown in FIG 2.4.

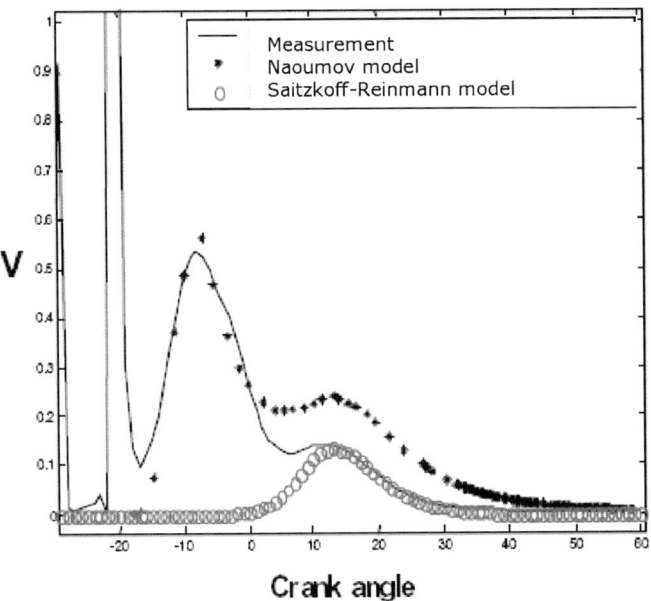

Figure 2.4   Comparison between Saitzkoff and Naoumov models [61].

In 2005, Mehresh and Souder [62] tried to simulate the ion current signal in an HCCI engine using detailed chemical kinetics for propane combustion, and included kinetics for ion formation. The ionization model contained 34 reactions and 9 ionic species NO^+, N^+, N_2^+, O_2^+, OH^+, O^+, H_3O^+, HCO^+, and electrons. Thermal ionization was not included in this model. They did not include any reactions of higher hydrocarbons with ions because their work was in the fairly lean regimes.

In 2007, Prager and Warnatz [63] modeled ionization in lean methane-oxygen flames involving an ion model containing 65 reversible reactions and 11 charged species, a set of positive and negative ions which are CHO^+, H_3O^+, CH_5O^+, $C_2H_3O^+$, O_2^-, OH^-, O^-, CO_3^-, CHO_2^-, CHO_3^-, and electrons. No engine work was done in this paper.

2.7 <u>CONCLUSION</u>

- From the literature review it is fairly clear that most of the researchers studied ion current in flames at atmospheric pressure because it is easy to collect ionic data using a mass spectrometer with direct contact to the flame. Eventually, some investigators started to work on ion current in metallic SI engines and create mathematical models to try to understand and correlate ion current traces obtained experimentally to the output of their models. All of the modeling work done in engines was conducted on spark ignited engines with homogeneous combustion where the equivalence ratio is fixed around unity.

- The famous models published in the last few years are basically the Staizkoff-Reinmann thermal model [5] which tried to explain and predicts only the second peak based on thermal ionization calculations of NO and Saha equation. They predicted the NO concentration and used it as an input to calculate the ion-current second peak. They had to bump up the NO concentration levels 3 to 10 times more than the actual one in order to get a match between their ion-current calculated and measured values. Increasing the [NO] level this way to reach a match means there is something missing and has not been accounted for in their model. Many people adopted that model later on and it has been used as for granted. Naoumov and Demin [61] published a better zero-dimensional chemical kinetic model, again for spark ignited engines, which can predict the first and the second peak based on chemi and thermal ionization.

CHAPTER 3   EXPERIMENTAL SETUP AND INSTRUMENTATION

3.1 CHAPTER OVERVIEW

Experiments were carried out in two engine test cells. The first test cell contains a John-Deere heavy duty commercial diesel engine. The second test cell contains a single cylinder optically accessible diesel engine designed only for research purposes. This chapter describes in details all Lab equipments used in conducting this research.

3.2 JOHN DEERE ENGINE

The engine used in part of this research is a 4.5L turbocharged, heavy duty, electronically controlled, direct injection commercial diesel engine provided by John-Deere. The engine is fitted with a hydraulic dynamometer and equipped with a common rail injection system, and six holes solenoid activated injectors. The water-cooled engine is equipped with a Variable Geometry Turbocharger (VGT) with electronic control system to regulate the intake pressure and maintain it at a desired level according to experimental requirements. Figure 3.1 shows an image of the engine including the VGT, fuel tank, dynamometer, radiator, and coupling cage used to secure the engine flywheel, rotating shaft and couplings. Engine specifications are listed in Table 3.1.

Figure 3.2 is an image of the cylinder head (Top Figure) and piston bowl (Bottom Figure). The cylinder head contains 4 valves, two for intake (Left side), and two for exhaust (Right side). The centrally located fuel injector contains six holes separated by 60° angle. The swirl motion direction, glow plug and pressure transducer locations are shown in the figure. In addition, the position of the glow plug hole in reference to the fuel injector jets is clearly represented. It is highly important to keep the glow plug, when used as an ion sensor, away from the fuel jets path as it affects the ion current detection. Figure 3.2 also shows the piston bowl reflecting the fuel jets traces. The swirl direction is marked on the shallow engine bowl.

TABLE 3.1

Engine Model	4045HF485		
No. of Cylinders	4	Length (mm)	860
Displacement (L)	4.5	Width (mm)	612
Bore and Stroke (mm)	106 x 127	Height (mm)	1039
Compression Ratio	17.0 : 1.0	Weight (Kg)	491
Engine Type	4 stroke	No. of Valves	16

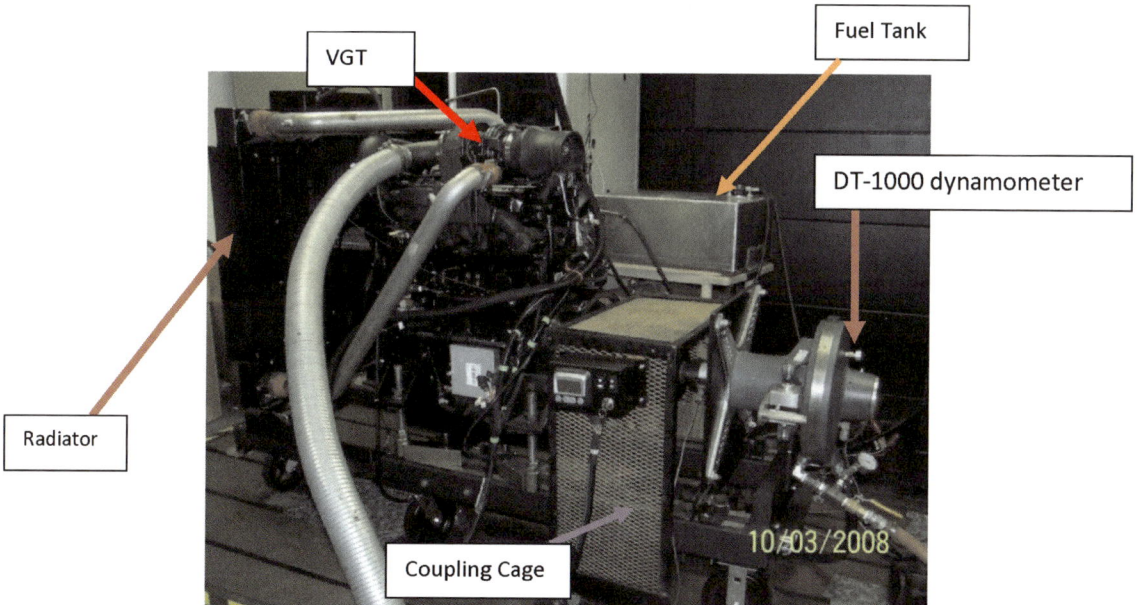

Figure 3.1 John-Deere direct injection diesel engine [64].

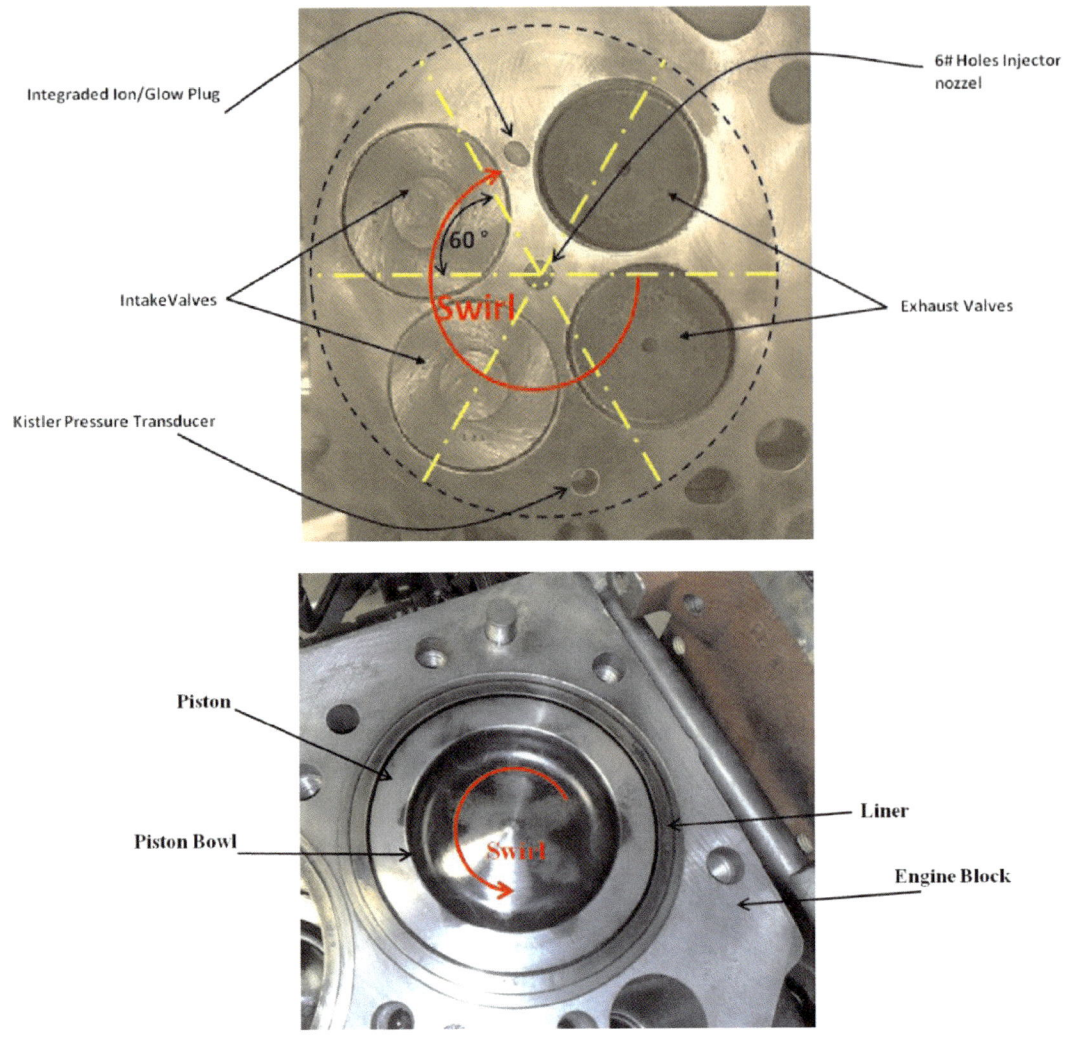

Figure 3.2   Cylinder head (Top Figure) and piston bowl (Bottom Figure) [64].

3.2.1 John Deere Engine Instrumentation

The John-Deere engine is heavily instrumented in order to meet experimental requirements. A Kistler piezoelectric pressure transducer is installed only in cylinder 1 to acquire the cylinder gas pressure signal. Furthermore, the fuel injector of cylinder 1 is equipped with a needle lift sensor to measure the needle plunger displacement. The engine is also equipped with a Kistler piezoresistive high pressure sensor to record the fuel pressure in the common rail system. An Omega pressure transducer is fitted in the intake manifold to monitor engine's intake pressure. In addition, intake temperature is recorded with a K-type thermocouple. Moreover, a fuel measuring system is used to keep track of the fuel consumption during engine operation. The Coriolis mass flow meter is able to measure flow rates as low as 1 gm/min.

3.2.2 Emissions Measurements

NO$_x$ is measured on a cycle-by-cycle basis with a fast response NO analyzer (CLD 500) provided by Cambustion. An image of the analyzer is shown in FIG 3.3. A sampling probe is used to extract gases from inside the engine cylinder or from the engine exhaust system for NO measurements. The principle of operation is based on the Chemi-Luminescence-Detection (CLD). Introducing ozone to NO sampled by the analyzer creates a chemical reaction that emits light. This reaction is the basis for the CLD in which the photons produced are detected by a photo multiplier tube (PMT). The CLD output voltage is proportional to NO concentration [65].

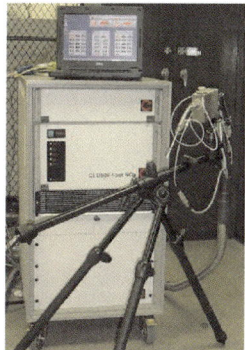

Figure 3.3   Fast NO analyzer (CLD500) provided by Cambustion [66].

An opacity meter is used to measure soot produced in the exhaust. The basic principle used in measuring smoke density is the attenuation of the intensity of a collimated light beam by smoke aerosol absorption and scattering from exhaust gases. Measurement is accomplished by passing light pulses through the engine exhaust stream and detecting the loss in light transmission due to exhaust smoke with a photoelectric detector. The relative light energy loss is translated into both opacity and smoke density signals, which are displayed digitally at the control unit and sent to the combustion analyzer by BNC cable [64]. Figure 3.4 shows the opacity meter mounted on the exhaust pipe of the John-Deere engine.

Figure 3.4   John-Deere engine equipped with an opacity meter [64].

3.3 <u>OPTICALLY ACCESSIBLE ENGINE</u>

An optically accessible direct injection 0.51L research diesel engine is used in the experimental work. The engine is provided by AVL Gratz, Austria. The single cylinder optical engine is fitted with a common rail fuel injection system rated to 1350 bar injection pressure. The solenoid activated fuel injector is centrally located inside the combustion chamber and has 5 symmetrically spaced holes of diameter 0.17 mm each. Engine specifications are listed in Table 3.2.

TABLE 3.2

Engine Model		AVL 5402	
No. of Cylinders	1	Con-Rod Length (mm)	148
Displacement (L)	0.51	Bowl Diameter(mm)	40
Bore and Stroke (mm)	85 x 90	Bowl Depth (mm)	17
Compression Ratio	15.0 : 1.0	Swirl Ratio	2.0 – 4.5
Engine Type	4 stroke	No. of Valves	4

Optical access is provided via an extended piston in which is mounted a 20 mm-thick fused-silica (quartz) window. The window is mounted within a metallic casing and sealed with an epoxy. An O-ring forms a seal between the piston and window casing, facilitating removal of the window/casing assembly during cleaning. The placement of a mirror inside the piston assembly at 45° relative to the cylinder axis provides a view of the entire 40 mm-diameter combustion bowl [67, 68].

The combustion bowl is of rectangular cross-section. Oil-less lubrication is provided by slotted graphite rings and the sealing by uninterrupted Bronze-Teflon rings. Compressed air directed at the underside of the piston window provides cooling of this assembly during operation [67, 68]. Figure 3.5 shows the extended piston and combustion bowl.

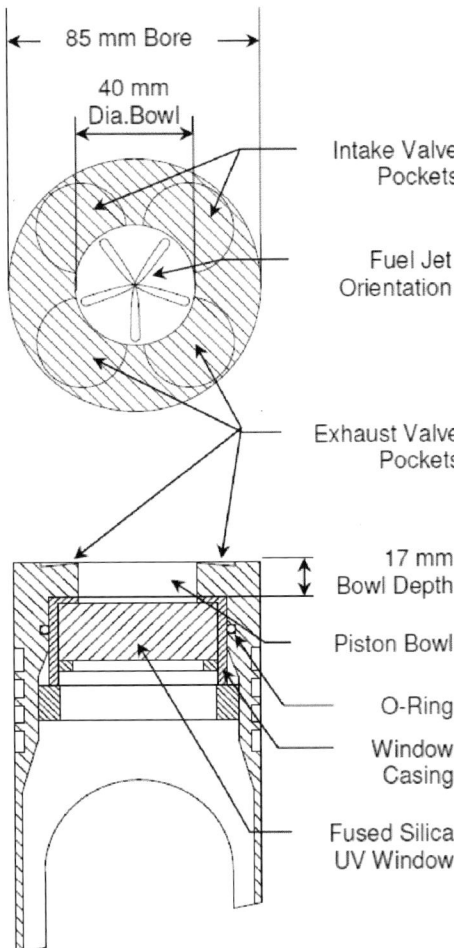

Figure 3.5   Extended piston and combustion bowl geometry [67, 68].

3.3.1 Optical Engine Instrumentation

The engine is driven by an AVL AC dynamometer at constant speeds. The engine is equipped with a non-cooled piezoelectric pressure transducer fitted in the cylinder head to measure cylinder gas pressure. An additional pressure transducer measures the pressure in the line between the fuel rail and injector. The fuel injector is instrumented with a needle lift detector to acquire the displacement of the needle plunger.

Images of combustion are recorded with the use of a high speed Phantom camera which can acquire up to 30 000 frames per second. The Phantom camera shown in FIG 3.6 is only capable of recording images in the visible range. However, an intensifier is needed to obtain shots of the combustion process in the Ultra-Violet (UV) range.

Figure 3.6   Phantom Camera used in recording combustion images.

CHAPTER 4 ION CURRENT MEASURING TECHNIQUES IN DIESEL ENGINES

4.1 <u>CHAPTER OVERVIEW</u>

The objective of this chapter is to discuss different ion current measuring techniques developed at Wayne State University in order to pursue my Ph.D research in diesel engines. Three ion probes were made to sense the ion-current signal in the engine cylinder. The first was a conventional glow plug modified to act as an ion sensor. This sensor requires a hole drilled in the cylinder head to accommodate the glow plug outer sleeve. I used this sensor to conduct experiments on the heavy duty John Deere metallic engine. The second sensor is called Multi-Sensing Fuel Injector (MSFI) where I use the fuel injector as an ion sensor. The MSFI can perform other sensing functions as well. I used the MSFI in the AVL single cylinder optically accessible diesel engine as the use of this sensor does not require a glow plug hole in the engine block. The third sensor is an In-Cylinder gas sampling probe modified to work as an ion sensor. This sensor was developed to study the correlation between certain sampled gases from the engine cylinder such as NO and the corresponding ion current in the same location. I used this sensor on the John Deere diesel engine where I fitted this sensor in the glow plug hole. The chapter will discuss the three measuring techniques in more details.

4.2 MODIFIED GLOW PLUG

Glavmo from Dephi, in 1999 was the first to modify a glow plug and use it as an ion current sensor [69]. The use of the modified glow plug as an ion sensor is limited to engines fitted with glow plugs. Other engines have to drill a hole in the cylinder head to accommodate a glow plug. At Wayne State University, I tried different designs to use the conventional glow plug as an ion sensor. The purpose of the design is to maintain the functionality of the glow plug while using it as an ion sensor. The cross section drawing and a picture of the new sensor is shown in FIG 4.1. The terminals of the glow plug circuit and ion current circuit are shown in the same figure. The original outer shell was used to maintain the distance of thread to the seat and the thread itself. The length of the inner sleeve was maintained same as the original design to avoid it hitting the piston head.

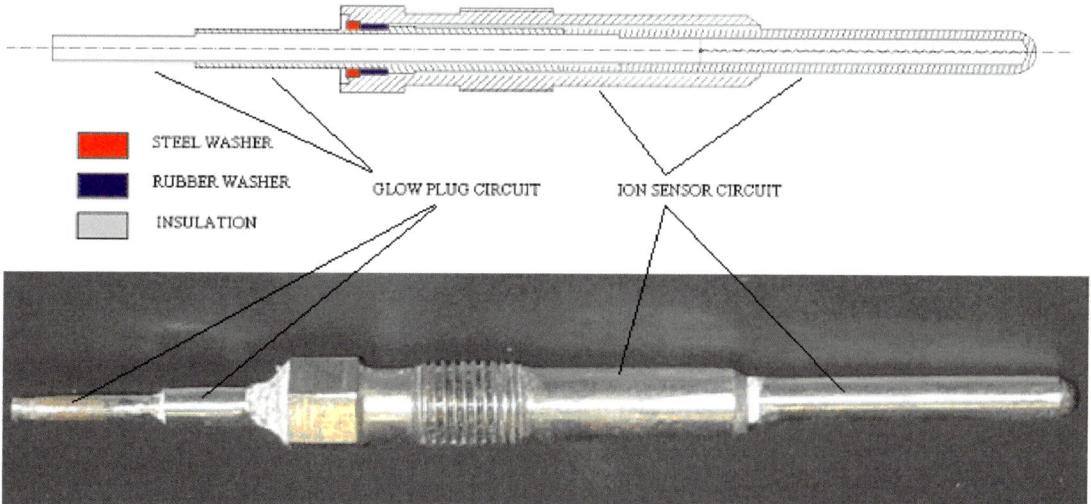

Figure 4.1 Convention glow plug / Ion-sensor designed at Wayne State University [66].

4.2.1 <u>Ion Current Circuit</u>

The ion current circuit comprises the modified glow plug, DC supply voltage, resistance and a signal conditioning system as shown in FIG 4.2. The inner rod of the modified glow plug is isolated from the engine body and connected to the positive terminal of the power supply. The outer sleeve of the glow plug is connected to the engine body to the negative terminal of the power supply. The inner rod of the glow plug is isolated from the outer sleeve using high temperature ceramic.

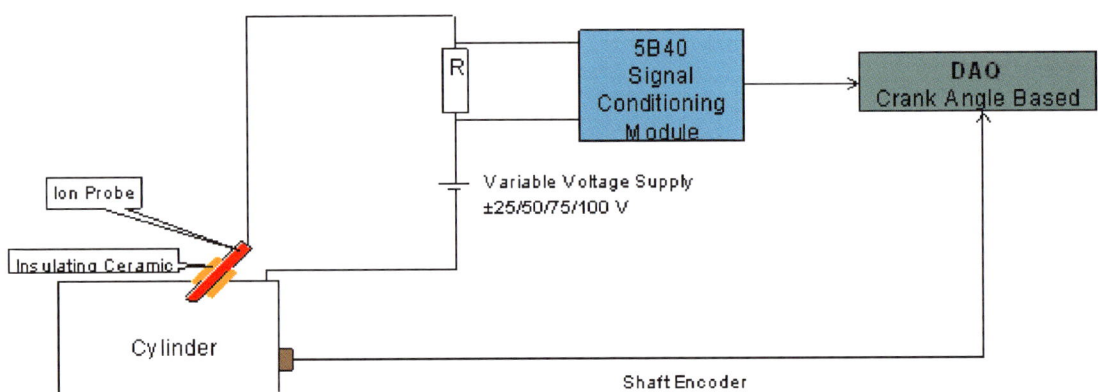

Figure 4.2 Ion current circuit using a modified glow plug as an ion sensor [66].

The DC supply voltage can produce 25, 50, 75, 100 volts. The measured current is of the range of micro amperes. A signal conditioning unit with 5B40 modules is used to amplify the ion current signal. These modules work at 10 kHz bandwidth and hence the phase delay is small. The input voltage range is ±100 mV and the output ±5 volt which means they amplify the signal 50 times. Figure 4.3 shows an image of the ion current circuit used in this dissertation.

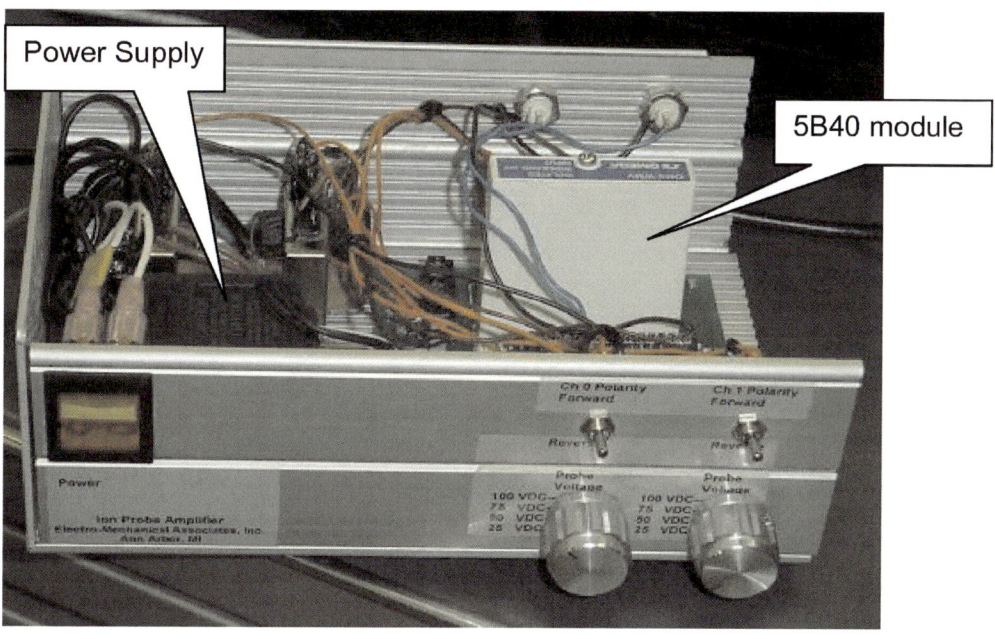

Figure 4.3   Signal Conditioning System and Power Supply [66].

4.3 <u>MULTI-SENSING FUEL INJECTOR</u>

The idea of using the fuel injector with electrical positive tip as an ion current sensor was first proposed by Volvo in 2003 [70]. However, at Wayne State University this idea was enhanced and implemented in such a way that the fuel injector is not only utilized as a current sensor, but also as a multi sensing device, referred to as "Multi-Sensing Fuel Injector", (MSFI) [71]. Its sensing ability is based on the Hall Effect and the ionization property of flames. The Hall Effect enables the MSFI to act as a current probe to detect the electric pulses for the start and end of injection. The ionization property of flames enables the MSFI to act as an ion current sensor. Using this technology, the fuel injector is able to perform five functions in addition to its main task of injecting fuel inside the combustion chamber. The fuel injector is capable of detecting fuel injection timing, combustion timing, and in-cylinder emissions. Furthermore, the MSFI can detect malfunctions over the lifetime of the engine such as injector fuel leakage and dribbling, improper injector driver operation, and engine misfire. The signal developed from the injector circuit is fed to engine ECU to control different engine operating parameters. The MSFI technology can be applied in direct injection diesel and gasoline engines operating in their conventional mode and in the HCCI regime. Furthermore, it can be retrofitted to already produced direct injection diesel and gasoline engines and this is why I used the MSFI on the optically accessible diesel engine.

A test was performed in the optically accessible diesel engine using the MSFI technology. Figure 4.4 shows the results obtained from the MSFI together with needle lift, rate of heat release and cylinder gas pressure traces. It demonstrates the periods when the injector acts as a Hall Effect device, and as an ion current sensor. Compared to spark plugs and glow plugs modified to work as ion sensors, the MSFI is the only sensor capable of combining two principals of operation in one signal as shown in FIG 4.4. The details of traces in this figure will be explained in the following sections.

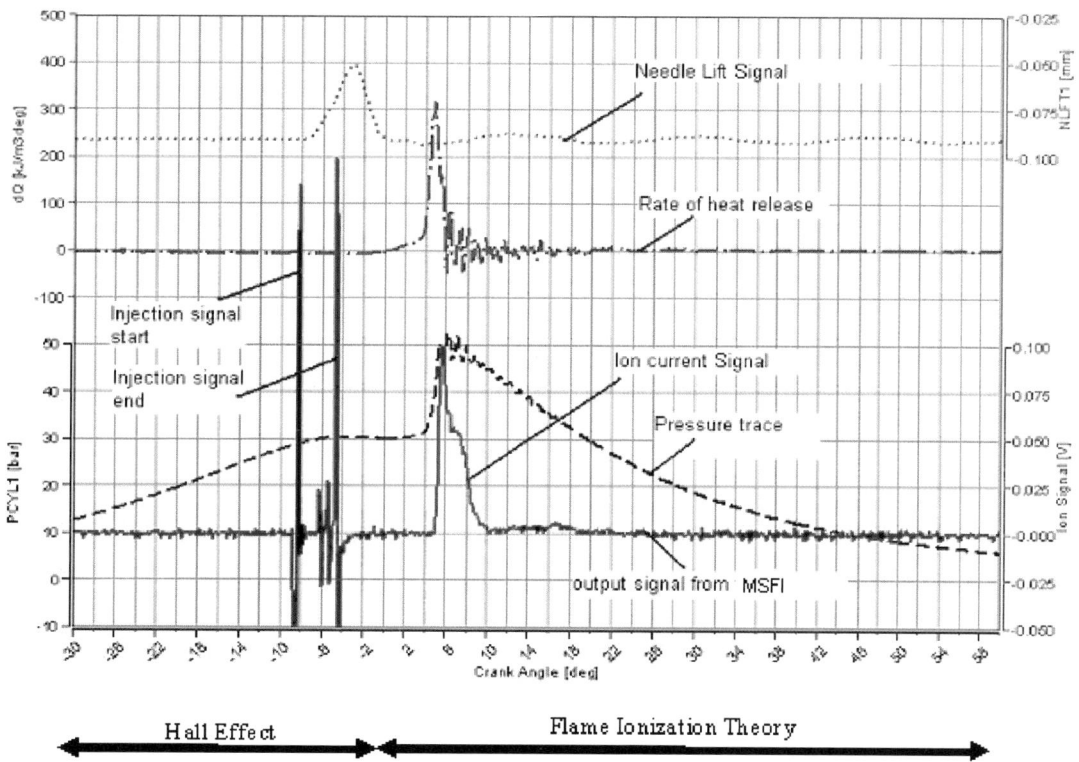

Figure 4.4   MSFI signal, cylinder gas pressure, needle lift, and Rate of Heat Release.
[Optical engine, rpm =1000, Start of Injection: 8.25°bTDC, Injection pressure: 400 bar]

4.3.1 MSFI Circuit

Figure 4.5 shows a schematic sketch of the MSFI system. The system consists of a pair of electrically isolated electrodes separated by a gap where the fuel injector acts as one electrode and the engine body acts as the other electrode. The fuel injector, located within the combustion chamber of the diesel engine, is electrically insolated from the rest of the engine body using various electrically non-conducting material parts. The first is in the form of a washer placed underneath the fuel injector to insulate it from the engine cylinder head. The second is placed between the injector holder and the cylinder head. The third is in the form of a coupling applied to the high pressure line upstream the fuel injector. An ion current electric circuit is then connected to the fuel injector in order to produce a signal indicative of different functions of MSFI system. The detection circuit includes a 100V DC power source with positive potential terminal connected to the fuel injector. The negative potential terminal is connected to the engine body. Current is measured in terms of the voltage drop across a resistor. Since the current signal tends to be relatively weak, a signal conditioning unit and an amplifier are included in the circuit. The amplifier can be integrated with low pass, high pass filters to reshape the incoming sensor output signal for control purposes.

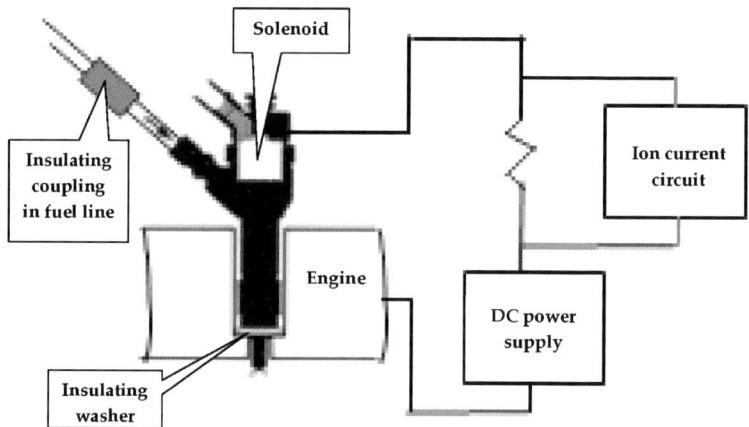

Figure 4.5   Insulation of a diesel engine fuel injector to act as MSFI and associated ion current circuit [72].

4.3.2 MSFI Sensing and Diagnostic Functions

4.3.2.1 Injection Timing

An experiment was carried out in a diesel engine equipped with a solenoid activated fuel injector to compare MSFI response to the injection pulse signal. In the arrangement shown in FIG 4.5, the injector is a part of two electric circuits. The first is the original circuit that includes the ECU and injector solenoid. The second is the ion current circuit. In order to determine the response of the second circuit to the pulse in the first circuit, a commercial current probe is clamped around the electric wiring to the injector solenoid. At the same time the MSFI signal is recorded. Figure 4.6 shows the output of the current probe and the corresponding MSFI signal. The electric pulse signal was sent to the solenoid at 8.5 CAD before TDC for a duration of 4 CAD. The ion current circuit signal showed two clear spikes with high amplitudes. The first spike coincides with the sharp rise in the current to activate the solenoid. The second spike coincides with sharp drop in the current to deactivate the solenoid. It should be noted that the two small spikes detected in the ion current signal at 5 and 6 CAD before TDC are caused by the change in the pulse current from a high level to a low level as shown in the current probe signal. From this analysis, it is clear that MSFI circuit can indicate the timing of both the activating and deactivating currents to the solenoid, based on the Hall Effect theory. In addition, the MSFI signal can detect the timing of the changes in the activation current from a high level to a low level.

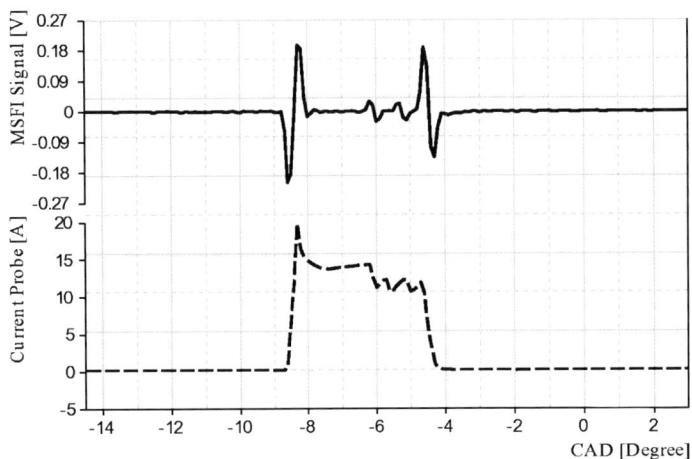

Figure 4.6   Comparison between MSFI and current probe signals produced by the electric pulse to the injector.

33

The signal produced by the MSFI in the optical diesel engine, given in FIG 4.7, shows the following:

a. The electric pulse to activate the injector solenoid and start the injection process is at 8.5°bTDC.

b. The electric pulse to deactivate the injector solenoid to end the injection process is at 4.5°bTDC. The period between the two signals, the injection pulse width (IPW), is 4 CADs.

c. Two low-amplitude signals between the signals stated in (a) and (b) are explained earlier, as being caused by the shift in the level of the current in the solenoid circuit from a high level to a low level.

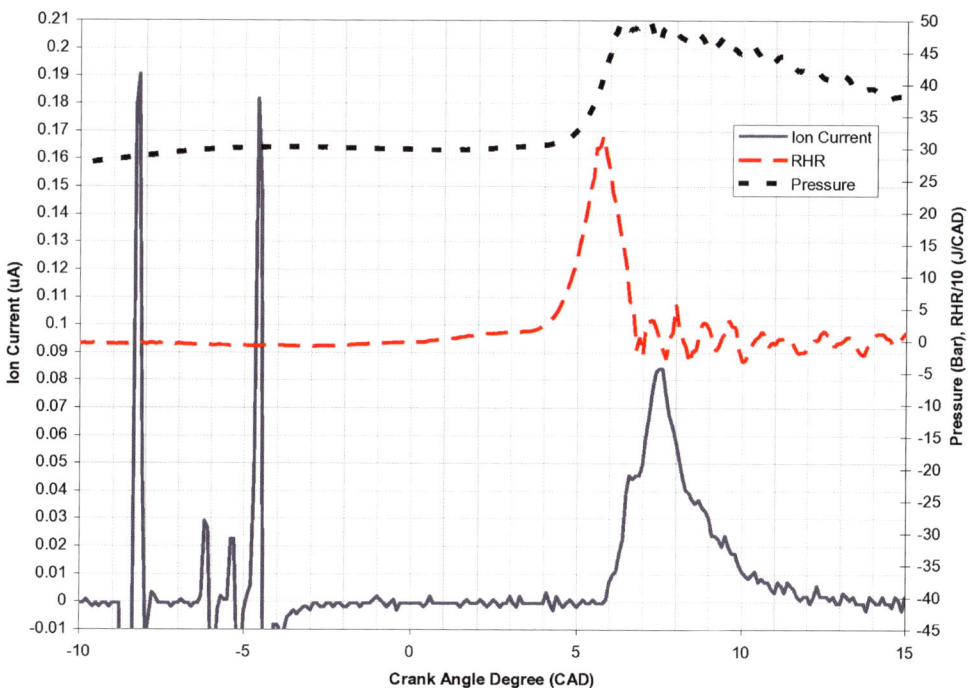

Figure 4.7 Traces of the ion current, cylinder gas pressure, and rate of heat release

4.3.2.2 Combustion Timing

Figure 4.7 shows the following:

a. The start of the rise in the ion current at 5.8° after TDC. This coincides with the location of peak of the rate of heat release due to premixed combustion, which coincides closely with the point at which the cylinder gas pressure reaches its highest rate of rise due to combustion.

b. The peak of the ion current, at 7.5° after TDC. This coincides with the end of the premixed combustion fraction.

c. The period from the activating current pulse spike to the start of combustion.

Another test was conducted in the optically accessible diesel engine where a high speed imaging PHANTOM camera is used to capture images of the combustion process at a speed of 30,000 frames per sec. The MSFI signal was also recorded as shown in FIG 4.8. The images show two combustion regimes. The first is the premixed combustion which is characterized by the blue flame. The second is the diffusion controlled combustion characterized by the sooty yellow flames. It is clear from the images that the start of the rise of the ion current, at 5.6 CAD after TDC, is associated with the peak of the rate of heat release in the premixed combustion regime.

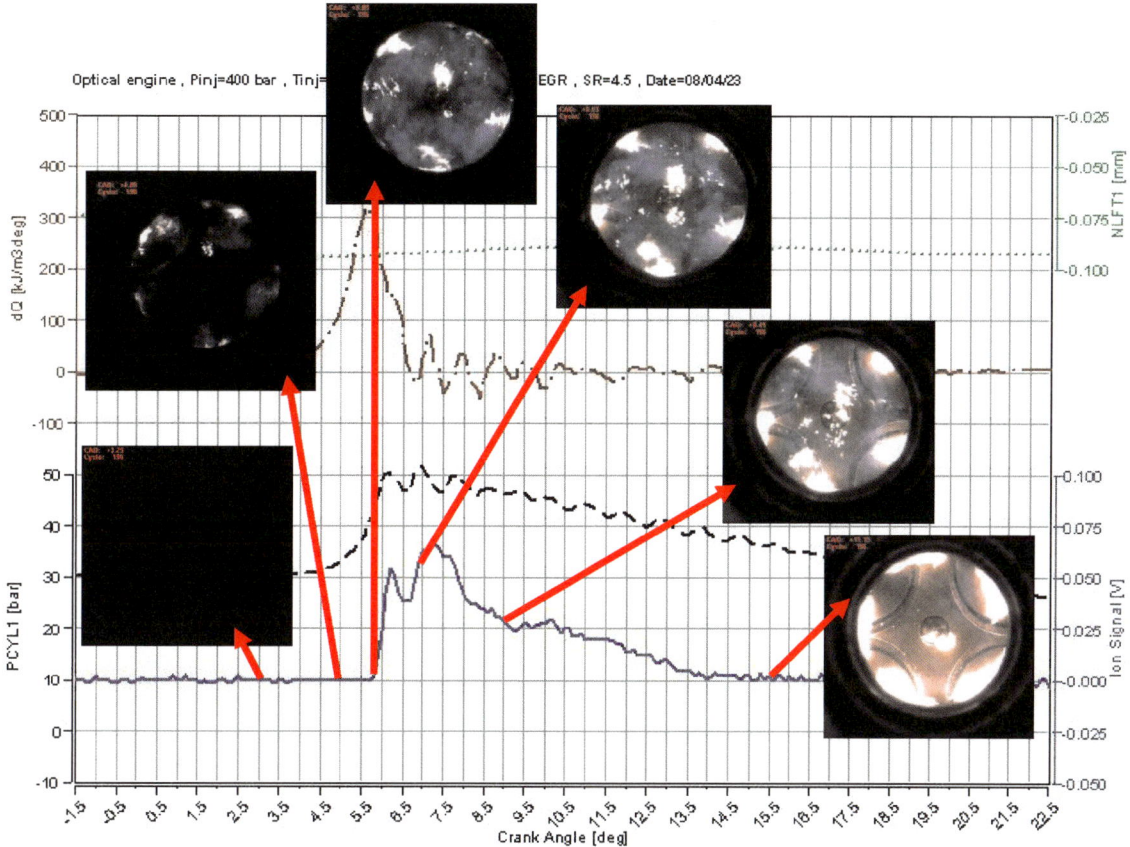

Figure 4.8   High speed camera images along with ion current measurement. Rate of Heat Release trace (dash-dotted red line) is shown on top followed by cylinder pressure trace (dashed black line) and ion current signal (solid blue line).

Another analysis has been carried out to study the effect of engine warm-up on the start of combustion location as well as start of ion current signal along the first 400 engine cycles of the optically accessible diesel engine.   Figure 4.9 reflects strong correlation between the location at which the ion-current signal starts and the first visible blue flame image caused by premixed combustion captured by the phantom camera. During cold start, late premixed combustion is caused by low engine temperature and thus longer ignition delay. Figure 4.9 reflects the start of combustion and ion current signal at 10 CAD after TDC in the first few cycles. As the engine warms up, ignition delay is shortened advancing the start of combustion and ion current until thermal stability is achieved.

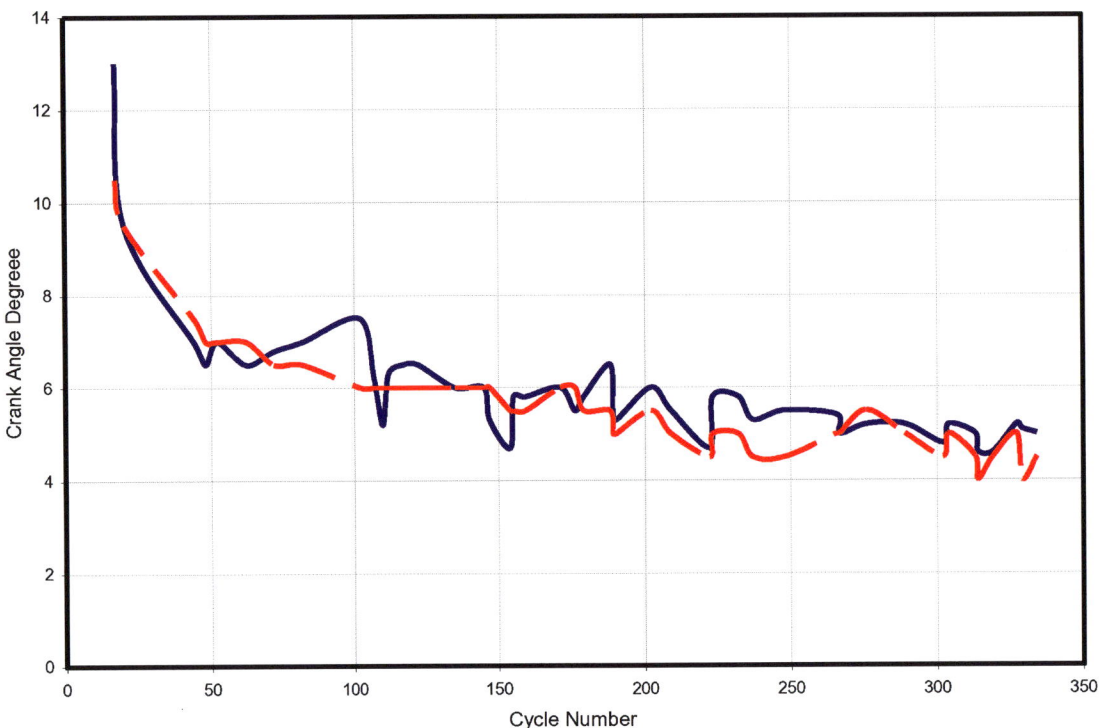

Figure 4.9   A correlation between the location of the start of ion current signal (solid blue line) and the first visible blue flame indicative of premixed combustion (dashed red line) during engine warm-up procedure.

4.3.2.3 <u>Injection Diagnostics</u>

The signal produced by the MSFI detected two malfunctions in the injection process, the first is fuel leakage from the injector and the second is a malfunction in the injector driver.

4.3.2.3.1 <u>Fuel Leakage</u>

Figure 4.10 demonstrates a cycle where fuel leakage from the injector was detected in the MSFI ion current signal. The traces are identical to those illustrated in FIG 4.4 except for a bump in the ion current signal that appears in the expansion stroke at around 20° after TDC. The duration and the peak of this bump indicate the presence of high concentrations of ions and electrons close to the injector tip. Out of 400 recorded engine cycles, 7 cycles were found with the late ionization current bump. The source of these ions was investigated by examining high speed images taken in the optical diesel engine at the same time the data in FIG 4.10 was recorded. The images for the cycles with ion bumps show a small luminous flame at the injector nozzle tip that coincides with the timing of these bumps. It is concluded that these bumps are caused by fuel leaking out of the injector holes, burning close to the surface of the injector tip and releasing charged particles captured by the MSFI.

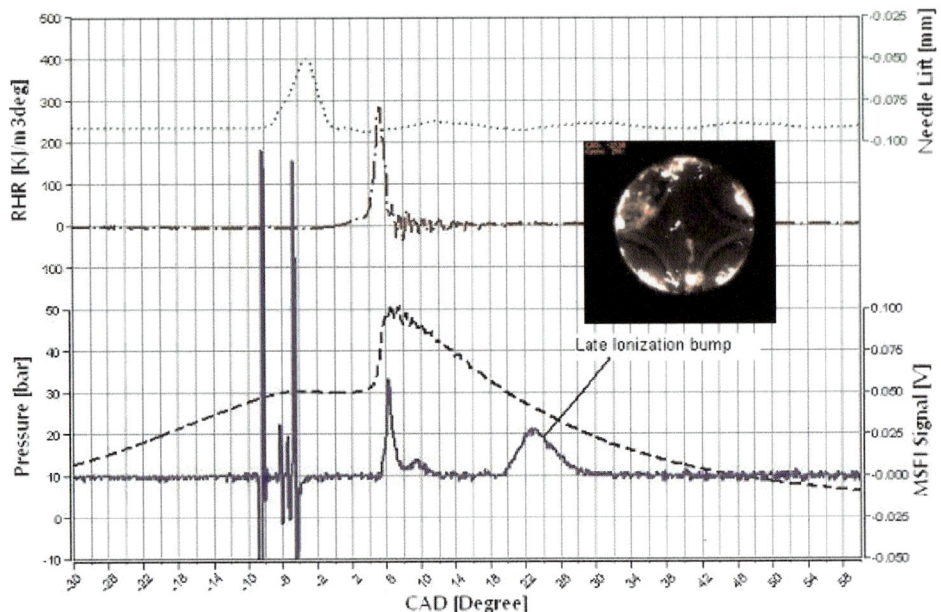

Figure 4.10   Output Signal from MSFI system with a leaking injector

4.3.2.3.2 Injector driver malfunction

Figure 4.11 shows MSFI signals for two cycles, and the corresponding traces for the needle lift and the electric current measured by a commercial current probe clamped around the wiring to the fuel injector. The first cycle is represented by solid lines, while the second cycle is represented by dotted lines. The preset amount of fuel delivery is the same for the two cycles. The second cycle shows a healthy operation, as the probe shows a sharp rise in the electric current to activate the solenoid to start fuel injection, and a sharp decay in the electric current to deactivate the solenoid to end the injection process. For this cycle, the MSFI signal shows two clear spikes of equal amplitudes that coincide with the current probe signals. The amount of fuel delivered in this cycle is reflected by the width and amplitude of the needle lift signal shown in the upper graph of FIG 4.11.

The first cycle represents a defective injector driver behavior that produced a higher lift and longer opening duration compared to the other cycle. The reason for such behavior can be found by analyzing the current probe signal. This signal shows a slower rate of decay in the deactivating current to the solenoid to close the needle. It took the current about 1.5 CAD more to drop in cycle 1 than in cycle 2 to deactivate the solenoid. A slow needle closing has serious implications on fuel atomization and penetration and ultimately on engine out emissions, particularly soot and hydrocarbon emissions. This malfunction of the driver can be detected from the MSFI signal, where the spike that corresponds to the needle closing shows much lower amplitude than that corresponding to the needle opening. This analysis demonstrates the ability of the MSFI to detect cycle-to-cycle variations in fuel injection.

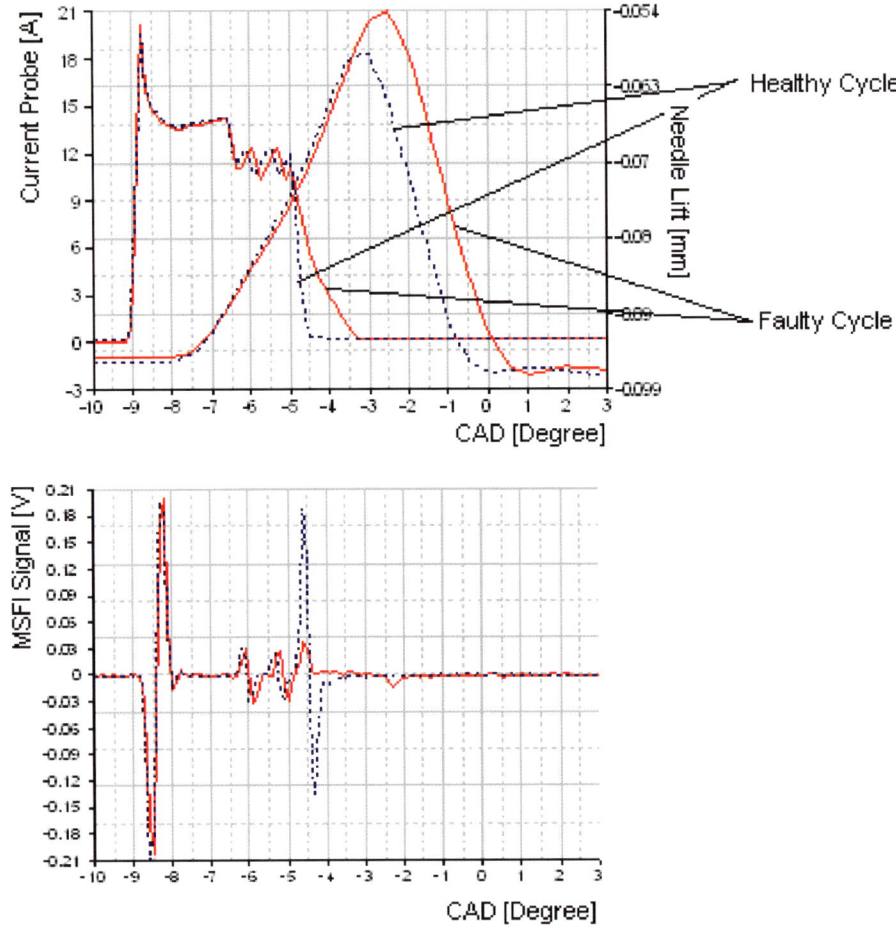

Figure 4.11   Comparison between two cycles representing a normal and a faulty injector driver.   The top figure shows the activating current measured by the current probe and needle lift. The bottom figure shows the output of the MSFI signal.

4.3.2.4 <u>Combustion Diagnostics</u>

The ion current signal depicts many characteristics of combustion in diesel engines. The MSFI being an integral part of each cylinder can produce a signal indicative of the state of combustion in the cylinder. This is particularly important in multi-cylinder engines if one cylinder is faulty. Any fault such as a misfire in one of the cylinders in a multi-cylinder engine can be a major source of loss of power, poor fuel economy and high hydrocarbon emissions.

A test was carried out using the optical diesel engine equipped with the MSFI system. Phantom camera images were recorded along with the MSFI signal. Figure 4.12 reflects cycle 20 during a cold start sequence where the engine is fairly cold with peak pressure of almost 38 bar. The weak MSFI signal reflects partially firing (close to misfiring) conditions as it is evident in the RHR and pressure traces. The corresponding images in FIG 4.12 show poor combustion with low light intensity.

Figure 4.13 represents cycle 210 in the cold start sequence and in this case the engine is fairly hot. The pressure trace has a peak value of almost 55 bar. Compared to FIG 4.12, the combustion in this cycle is successful with no evidence of misfire. This is reflected on the MSFI signal which is more significant. In addition, Phantom camera images in FIG 4.13 represent clearly the two modes of combustion (premixed and diffusion controlled).

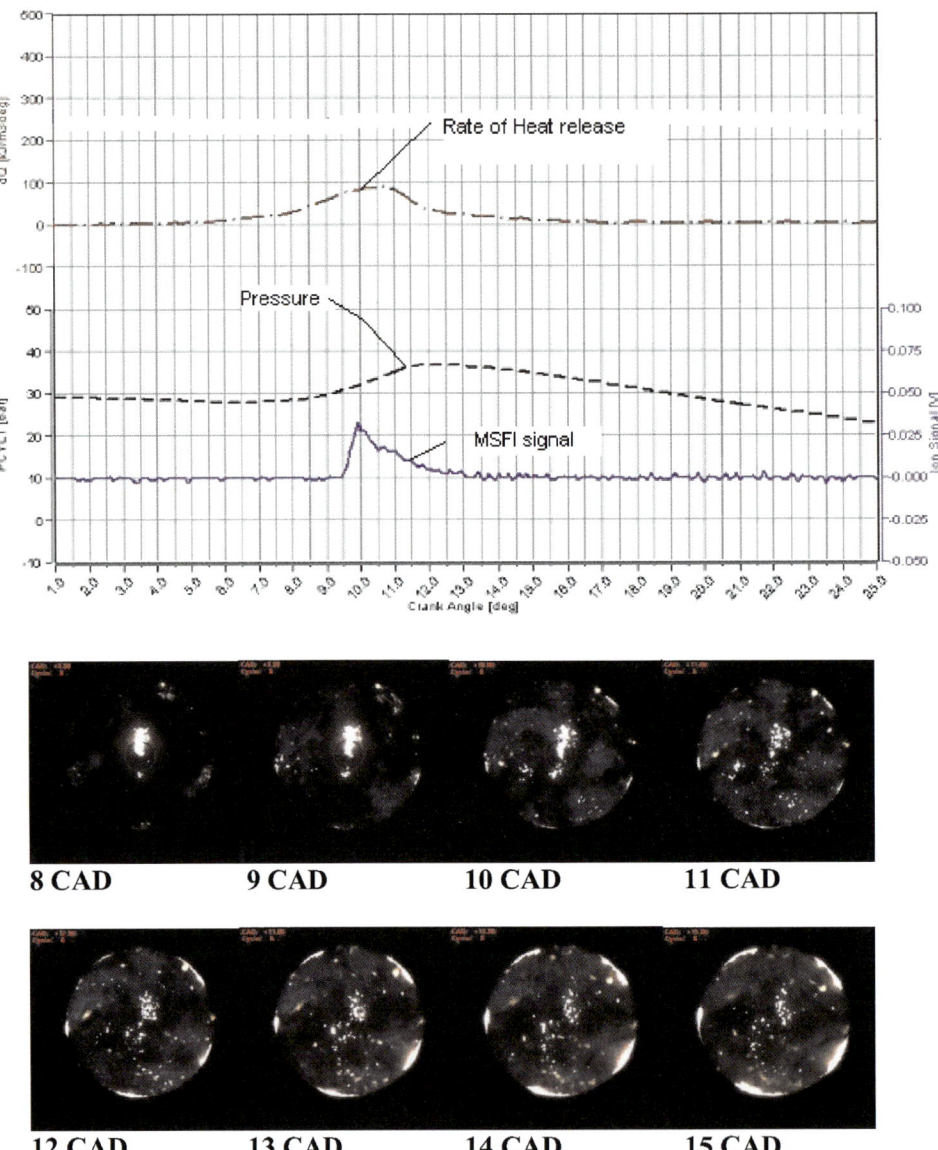

8 CAD   **9 CAD**   **10 CAD**   **11 CAD**

12 CAD   **13 CAD**   **14 CAD**   **15 CAD**

Figure 4.12   Cycle number 20 during an engine cold start experiment. High speed camera images along with MSFI measurement. Rate of Heat Release trace (dash-dotted red line) is shown followed by cylinder pressure trace (dashed black line) and MSFI signal (solid blue line).

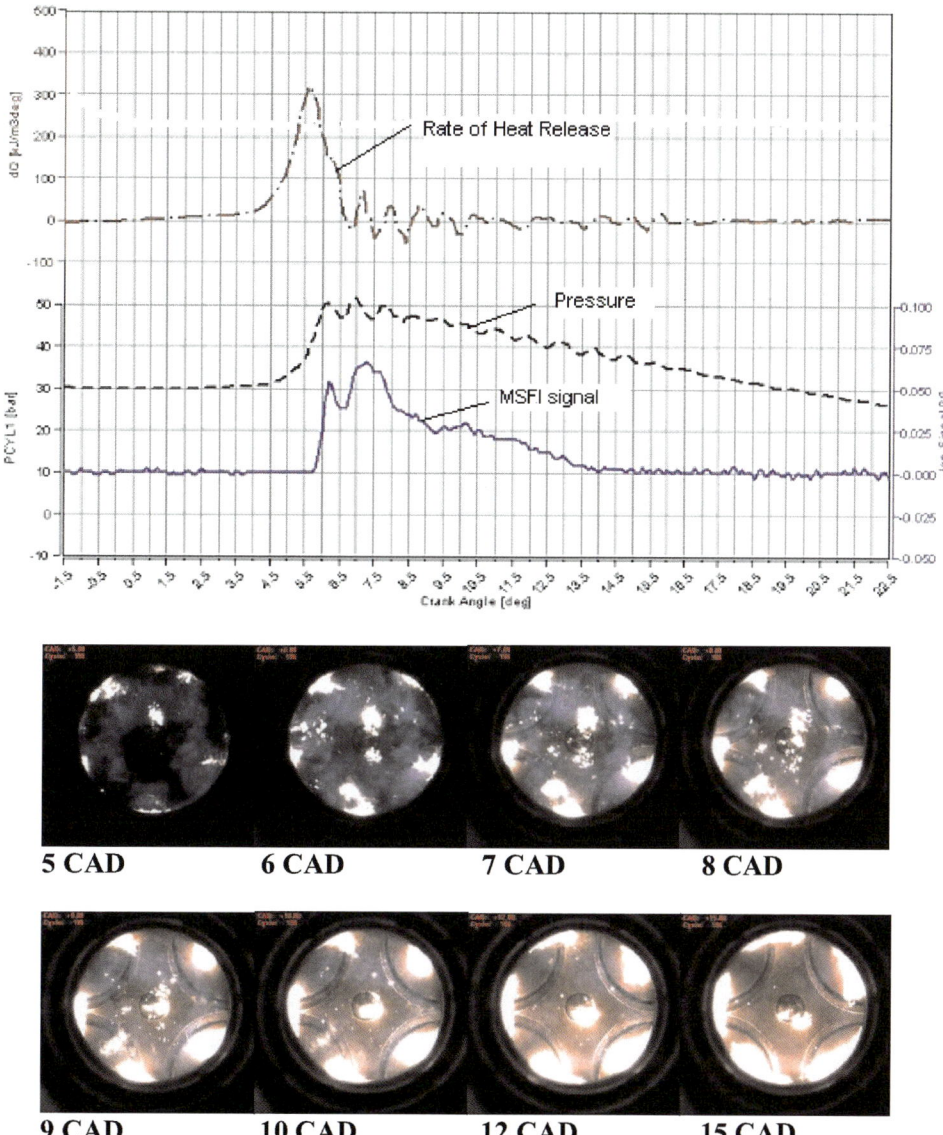

5 CAD   6 CAD   7 CAD   8 CAD

9 CAD   10 CAD   12 CAD   15 CAD

Figure 4.13   Cycle number 210 during an engine cold start experiment. High speed camera images along MSFI measurement. Rate of Heat Release trace (dash-dotted red line) is shown followed by cylinder pressure trace (dashed black line) and MSFI signal (solid blue line).

4.3.3 MSFI Control Algorithm

Figure 4.14 outlines a flow chart for an algorithm the ECU can follow to verify various operating conditions of the engine based on the MSFI signal [71, 72]. The ECU initially determines whether or not a start of injection spike (SOIS) is detected. If the ECU does not detect this spike (SOIS), then it concludes that the fuel injection system failed to deliver fuel into the combustion chamber. In a following step, if the ECU detects SOIS and does not detect the end of injection spike (EOIS), then it concludes an abnormality in the injector driver as well as the delivery of a larger amount of fuel than the preset value. If it detects EOIS, then it proceeds to the subsequent step.

In the next step, the ECU determines whether or not a start of combustion peak (SOCP) is detected. If no, this means that combustion did not occur. In one final step, if the ECU detects a fuel leakage bump or fuel dribbling peak (FDP) in the MSFI signal, the ECU concludes that combustion has occurred but with a fuel leakage problem. The ECU can make a decision about the fuel injector reliability, and whether or not it needs to be changed depending on the frequency of FDP peaks in the signal. If the FDP peak is not detected, then the ECU concludes that the injection of fuel and the combustion thereof are successful. Accordingly, the output signal of the MSFI can be used as a feedback signal to the ECU for engine control and diagnostics.

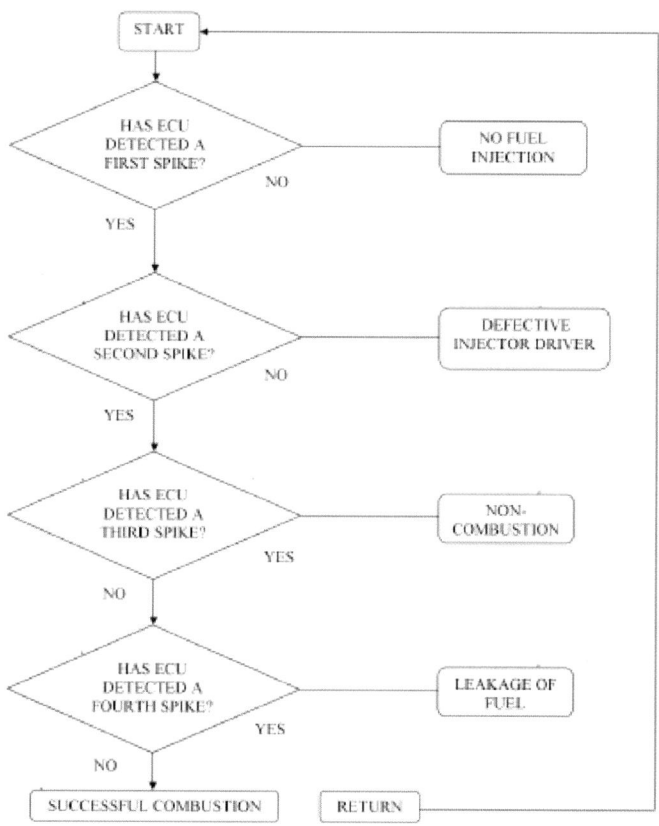

Figure 4.14   Flowchart illustrating an algorithm for ECU to follow using MSFI signal.

4.3.4 Mathematical Model to Compare Between Conventional Glow Plug Ion-Sensor and MSFI

An investigation is carried out to compare between the electric field generated by a conventional ion probe such as a modified glow plug, and the centrally located MSFI. QuickField, finite elements analysis software, is used for the electric fields simulation [127]. This software has the ability to simulate uniform and non-uniform electric fields [128]. The strength of the electric field is determined by the voltage between the charged surfaces and the distance separating them. Uniform electric fields are created between two parallel charged plates, concentric spheres, or coaxial tubes. The electric field in the combustion chamber is not uniform, more complex and depends on the location of the ion probe.

In this investigation, a charged particle (electron or ion), is located inside the combustion chamber having a non-uniform electric field existing between the ion probe (positively charged) and the engine body (ground). The potential difference is 300 V in this case. The ion probe is the glow plug in one case and the MSFI in the second. Figure 4.15 illustrates the simulation results. A charged particle is placed once in the right half of the engine bowl and once more in the left half of the bowl. The charged particle has the possibility to follow 360 different paths. Six paths were chosen every 60 degree.

Figures (4.15a) and (4.15b) show the simulated charged particle paths (in red lines) to a glow plug used as an ion sensor. Figure (4.15a) shows the charged particle on the left side was able to reach close to the plug. However, the particle on the other

47

side of the bowl shown in figure (4.15b), could not reach the plug because of the long distance where electric field is weak. Therefore, only the charged particles located on one side of the combustion chamber have the chance to reach the glow plug. This confirms the statement that the glow plug on any other ion current probe which is located on one side of the combustion chamber is a local sensor [73- 75].

On the other hand, figures (4.15c) and (4.15d) simulate the paths of the charged particles to the centrally located MSFI. Both figures show an electric field that covers the whole combustion chamber with uniform strength. Accordingly, it can be stated that the MSFI has an advantage over the conventional ion probes, since it does not act as a local sensor.

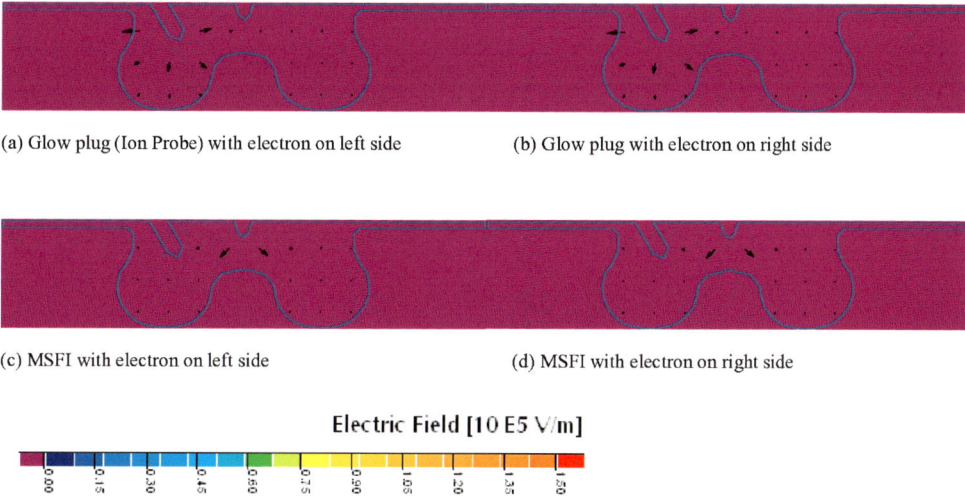

(a) Glow plug (Ion Probe) with electron on left side   (b) Glow plug with electron on right side

(c) MSFI with electron on left side   (d) MSFI with electron on right side

Electric Field [10 E5 V/m]

Figure 4.15   Simulation results comparing between the electric field generated by a conventional glow plug ion sensor and the new MSFI.

4.4 IN-CYLINDER GAS SAMPLING/ION-CURRENT PROBE

Currently, there is no sensor that determines the correlation between chemical products developed during combustion and ion current signal. Combining in-cylinder sampling techniques with ion current detection is of great importance as it is necessary to better understand how ionization processes are tied to different combustion products, and thus create new methods to control internal combustion engines under various operating modes and fuels. Ion-sensors are local sensors [73- 75], they can only measure current created by species ionized in the vicinity of the sensor. Therefore, in-cylinder sampling probes and ion sensors have to measure their signals at the same exact location in order to study the effect of different sampled species on the ionization signal.

In this dissertation, modifications to an existing in-cylinder sampling probe have been made to accommodate ion current measurements. The new In-Cylinder gas sampling/Ion-current probe has been patented through Wayne State University [76]. With the use of this novel sensor, I was able to study and reveal the effect of various specific species on the ion current signal as far as amplitude and shape. This sensor can accurately determine the contribution of the in-cylinder sampled species on the ionization signal. The outcome of the use of this new probe will be shown and discussed in details in Chapter 6.

4.4.1 <u>Sensor Design</u>

The In-Cylinder gas sampling probe was modified to sense ion current as discussed above. The top image of FIG 4.16 shows an original gas sampling probe purchased from Cambustion, a company located in England, to measure In-Cylinder NO using a Fast NOx machine. The bottom image of FIG 4.16 shows the modifications made at Wayne State University to adapt the gas sampling probe for simultaneous measurement of ion current. A ceramic high temperature white coating is used to cover and insulate the sampling pipe electrically from the rest of the engine body. A glow plug adapter is then used as an outer sleeve to fit the sampling probe in the glow plug hole of the engine cylinder head.

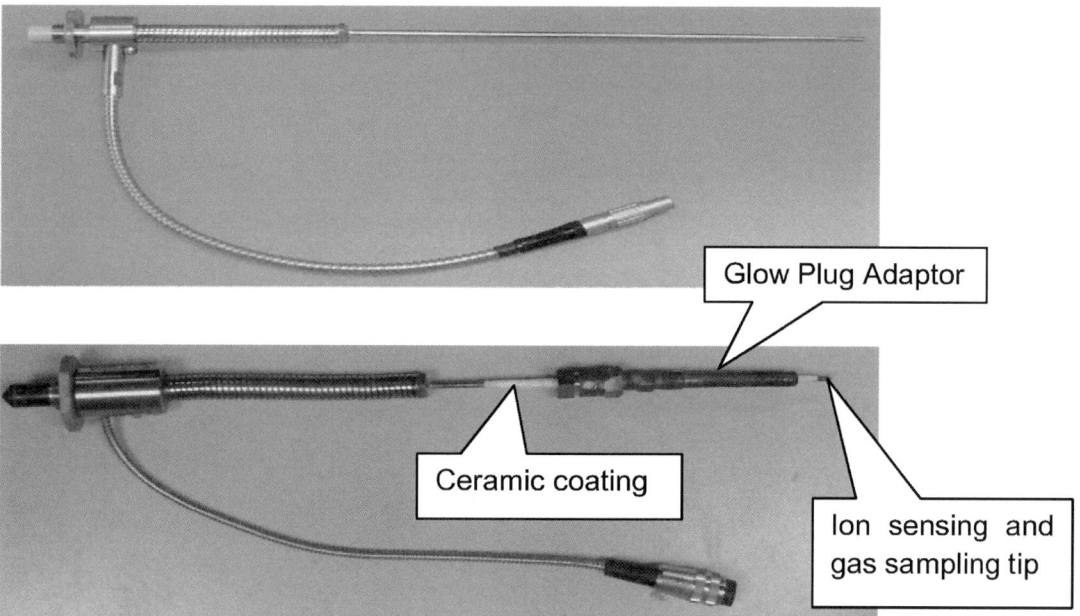

Figure 4.16 Top image – Original Gas sampling probe purchased from Cambustion, while Bottom image – Gas sampling probe modified to sense ion current.

4.4.2 Ion Circuit

After modifying the sampling probe to accommodate ion sensing, an ionization circuit is then added. Figure 4.17 represents a detailed outline of the ion current circuit used to operate the sensor. A DC power supply with preset positive potential is used. The in-cylinder sampling tube is connected to the positive terminal. The outer sleeve (glow plug adapter) is connected as well as the engine body to the negative terminal of the power supply. Voltage drop is measured across a resistor and processed via a signal conditioning unit.

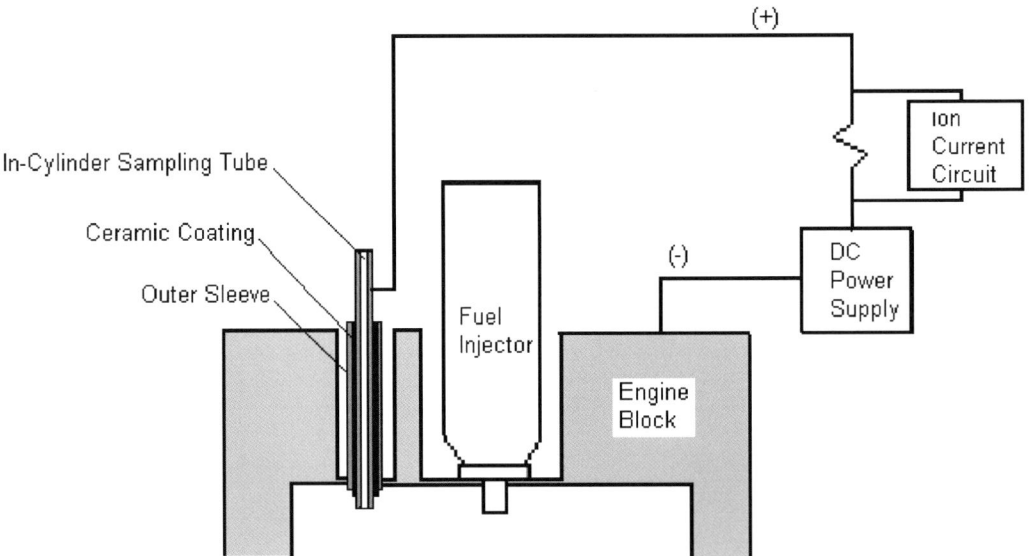

Figure 4.17   Ion current circuit required for the new probe [76].

4.4.3 <u>System Hardware</u>

Figure 4.18 outlines the arrangements required for full operation of the sensor in order to extract gases from the engine cylinder and to sense ion current. Two paths are shown in the figure, one is for gas sampling and the other is for ion sensing. First, the sampling tube is fitted inside the engine cylinder through the glow plug hole. Gas is sampled via the probe onto a Cambustion gas analyzer (CLD 500) and then to a data acquisition system. Second, Ion current is measured at the tip of the gas sampling probe using the ionization circuit discussed above and then to the same data acquisition system.

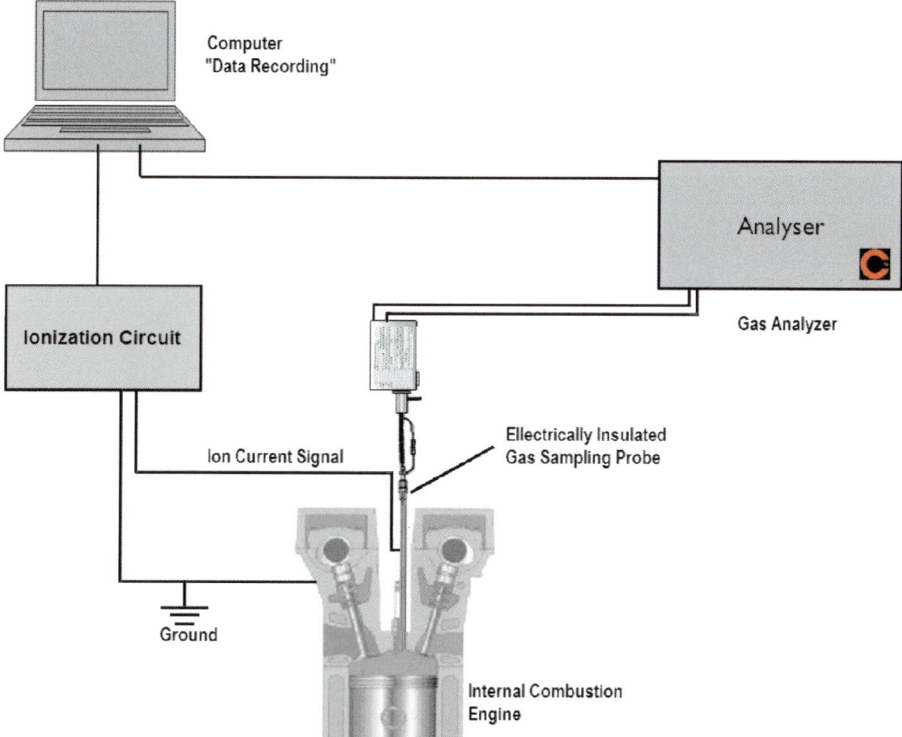

Figure 4.18   Gas sampling system equipped with ion current circuit.

4.5 <u>CONCLUSION</u>

- The conventional ion current sensor (modified glow plug) was first used to measure the ion current signal during engine tests. However, the glow plug was not suitable for all of the experiments I designed to complete my thesis. Therefore, new ion measuring techniques were developed and implemented such as a Multi-Sensing Fuel injector (MSFI) and an In-Cylinder Gas Sampling / Ion-Current probe.

- The MSFI is not only used as an ion sensor, but also as a multi sensor. It is able to act as an injection timing sensor, and an injection and combustion diagnostic tool. Furthermore, the signal produced by the MSFI is an indication of possible malfunctions in the injection system such as fuel leakage from the injector holes and abnormalities in the injector driver operation.

- A control algorithm has been developed for the use of the MSFI output signal in the feedback loop to the engine ECU for injection timing and injection and combustion diagnostics.

- An in-cylinder gas sampling probe was modified to measure the ion current signal and this new measuring technique allows gas sampling and ion current measurement to be performed at the same exact location. This is very important as we will be able to study the effect of certain species (the sampled gas) on the ion current behavior.

CHAPTER 5   SOOT AND ION CURRENT CORRELATIONS

5.1 CHAPTER OVERVIEW

The objective of this chapter is to discuss the correlation between soot formed inside the combustion chamber and the ion current. Experimental work was conducted on the heavy duty John Deere diesel engine under transient and steady state operating conditions using open and closed ECUs. A conventional glow plug modified to sense ion current was used during this investigation. Detailed analysis was carried out reflecting positive soot contribution to the ion current signal.

Based on the analysis given in this chapter, a new technique has been developed at Wayne State University and patented [88] to determine the soot content in the cylinder from the ion current signal. The soot content determined by this technique is an indicator of the soot concentration in the exhaust. One of the main advantages of this technique is that it shows the soot content on a cycle-by-cycle basis under different steady and transient engine operating conditions. In addition, the soot content in each cylinder can be continuously detected by the ECU. Any malfunctioning cylinder could be easily identified by the ECU and corrective actions can be taken for this specific cylinder.

5.2 EXPERIMENTAL SETUP

The experiments throughout this chapter were conducted on a multi-cylinder heavy duty John Deere diesel engine where cylinder 1 was equipped with a pressure transducer and a conventional glow plug modified to work as an ion current sensor. The engine is equipped with a common rail injection system and variable geometry turbocharger (VGT). An opacity meter is fitted on the exhaust pipe for soot measurements as shown in FIG 5.1.

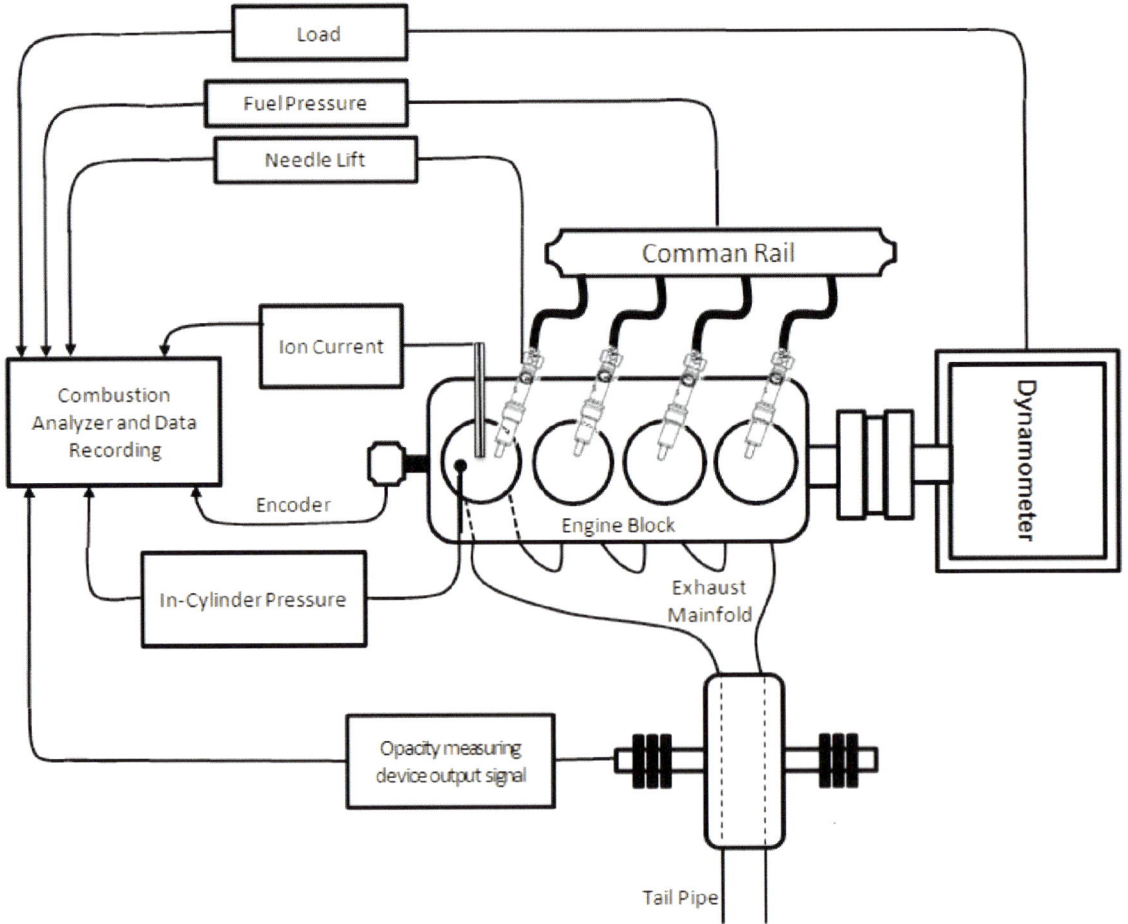

Figure 5.1   Experimental layout used for soot and ion current measurements.

A test matrix, designed to study the contribution of soot in the ion current signal, is shown in Table 5.1. The engine was controlled using an open engine control unit (ECU), where engine parameters such as injection timing, injection pressure, intake pressure, and engine speed were controlled. The engine test was performed at a low injection pressure of 400 bar as well as at a high injection pressure of 1000 bar. A sweep of loads of 5, 7, 9, and 11 bar IMEP was conducted at each injection pressure. Start of fuel injection is kept the same in all tests. Speed is maintained constant at 1300 RPM. In addition, the Variable Geometry Turbocharger (VGT) was used to maintain the same intake pressure during tests. Figure 5.2 shows cylinder pressure traces, needle lift signal, calculated rate of heat release, and cylinder temperature at 400 bar (top graph) and 1000 bar (bottom graph) injection pressure .

TABLE 5.1

Injection Pressure	400 bar		1000 bar	
Engine Load	SOI	Fuel Flow (g/min)	SOI	Fuel Flow (g/min)
5 bar	8 bTDC	82.3	8 bTDC	75
7 bar	8 bTDC	117	8 bTDC	108
9 bar	8 bTDC	163	8 bTDC	131
11 bar	8 bTDC	197	8 bTDC	173

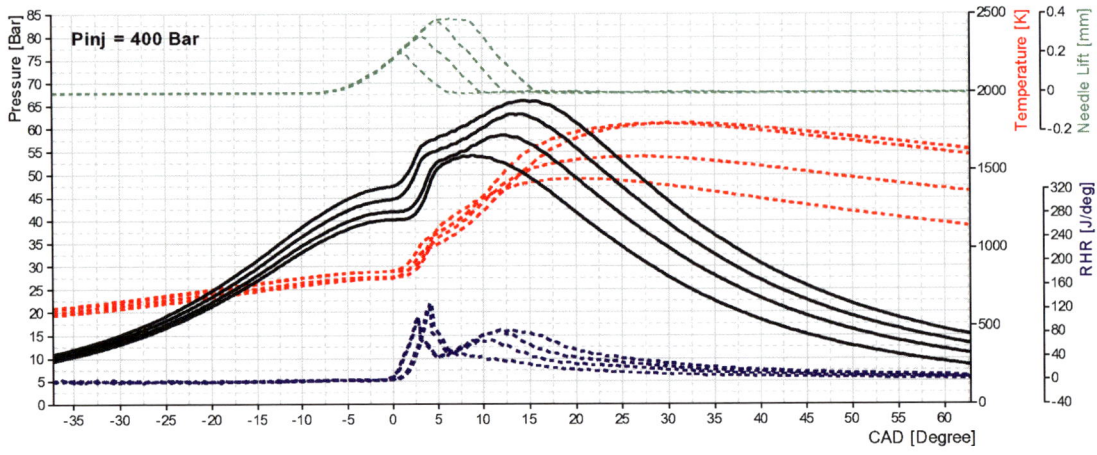

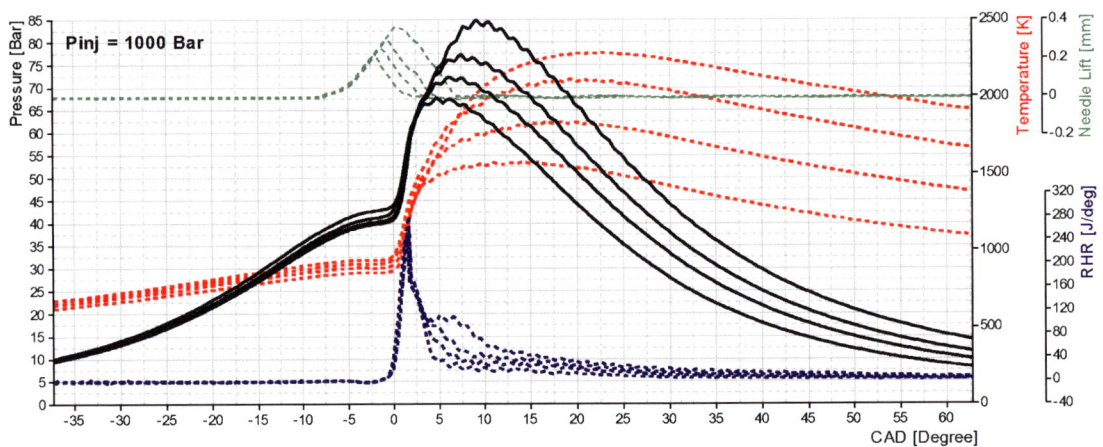

Figure 5.2   Recorded cylinder pressure traces (Black) and needle lift signal (Green), and calculated rate of heat release (Blue) and cylinder temperature (Red) for a sweep of engine loads at two injection pressures, 400 bar (Top figure) and 1000 bar (Bottom figure).

5.3 <u>CYCLE-BY-CYCLE ION CURRENT SIGNAL ANALYSIS</u>

As a first step, I tried to study the correlation between the ion current signal and the soot measured in the engine exhaust port on a cyclic basis. As shown in FIG 5.3, the maximum ion current amplitude in each cycle is plotted versus soot. The figure consists of 4 graphs representing lowest (5 bar IMEP) and highest (11 bar IMEP) loads at each injection pressure.

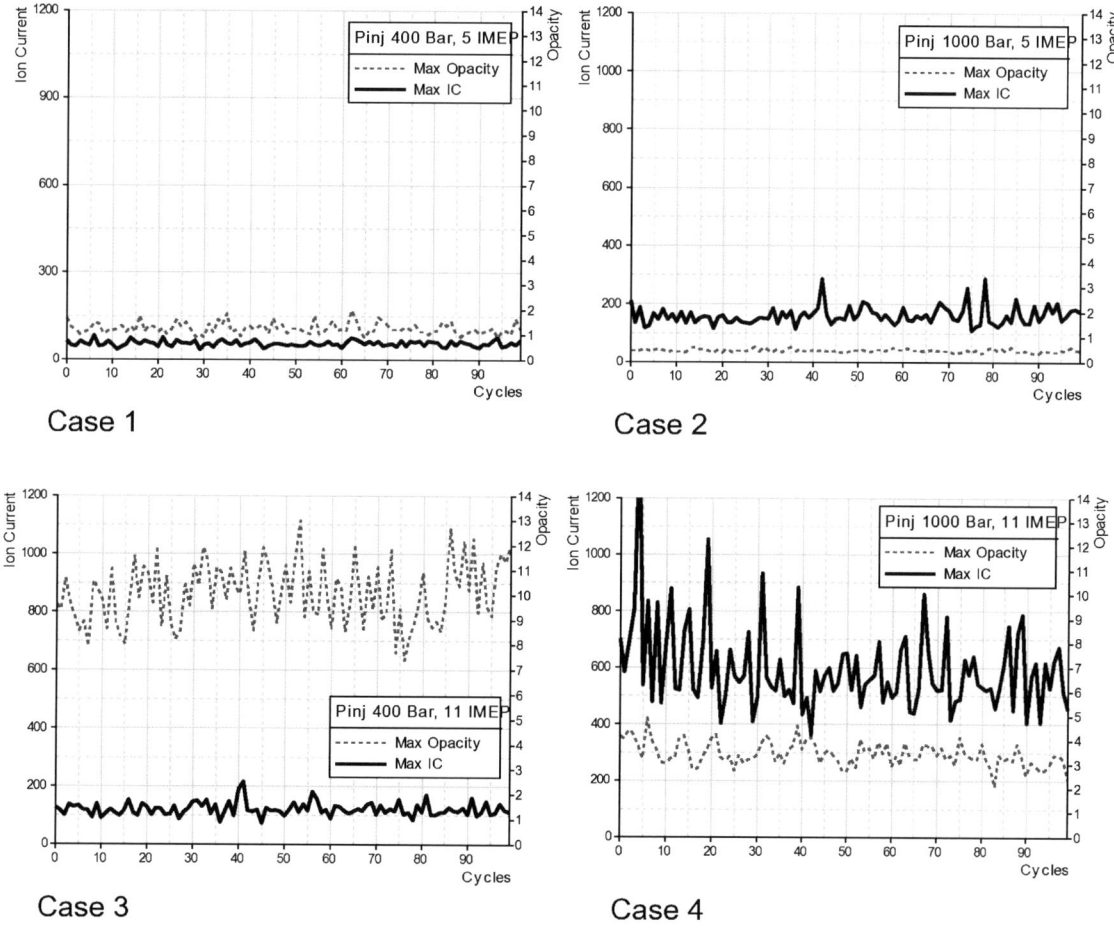

Figure 5.3   A set of results showing ion current maximum amplitude (Max IC) and maximum soot (Max Opacity) measured on a cyclic basis.

- (CASE 1) <u>Injection Pressure 400 bar, and 5 IMEP load</u>: As shown in the upper left graph of FIG 5.3, the ion current peak fluctuates around 60 uA and soot percent is close to 1.5.

- (CASE 2) <u>Injection Pressure 1000 bar, and 5 IMEP load</u>: As shown in the upper right graph of FIG 5.3, the ion current peak fluctuates around 180 uA and soot percent is close to 0.4.

- (CASE 3) <u>Injection Pressure 400 bar, and 11 IMEP load</u>: As shown in the lower left graph of FIG 5.3, the ion current peak fluctuates around 120 uA while soot reached its highest levels close to 11 %.

- (CASE 4) <u>Injection Pressure 1000 bar, and 11 IMEP load</u>: As shown in the lower right graph of FIG 5.3, the ion current peak fluctuates around 600 uA and soot percent is close to 3.5.

The analysis shows that the ion current peak reached its lowest levels at low loads and low injection pressures. It increased slightly when the engine ran at higher loads. On the other hand, at high injection pressure, the ion current amplitude increased dramatically reaching its highest levels by increasing engine load.   As far as soot signal, it is clear that soot produced at lower injection pressure is fairly high, especially at higher engine loads.

A more detailed analysis was carried out to study the relationship between the ion current peak and soot signal within each of the 4 cases listed above. The data in FIG 5.3 was re-plotted as shown in FIG 5.4 to come up with the correlation factor in each case. In FIG 5.4, maximum ion current is plotted versus maximum soot where each of the 4 graphs contains 100 points corresponding to 100 engine cycles. As shown, the correlation coefficients for cases 1, 2, 3, and 4 are 15.8%, 8.7%, 14.7%, and 29.8% respectively. These percentages reflect a fairly poor correlation between the ion current peak and maximum soot produced in all 4 cases. Thus the ion current peak cannot be used as a representative of soot, however this fact does not conclude that soot does not contribute to the ion current signal.

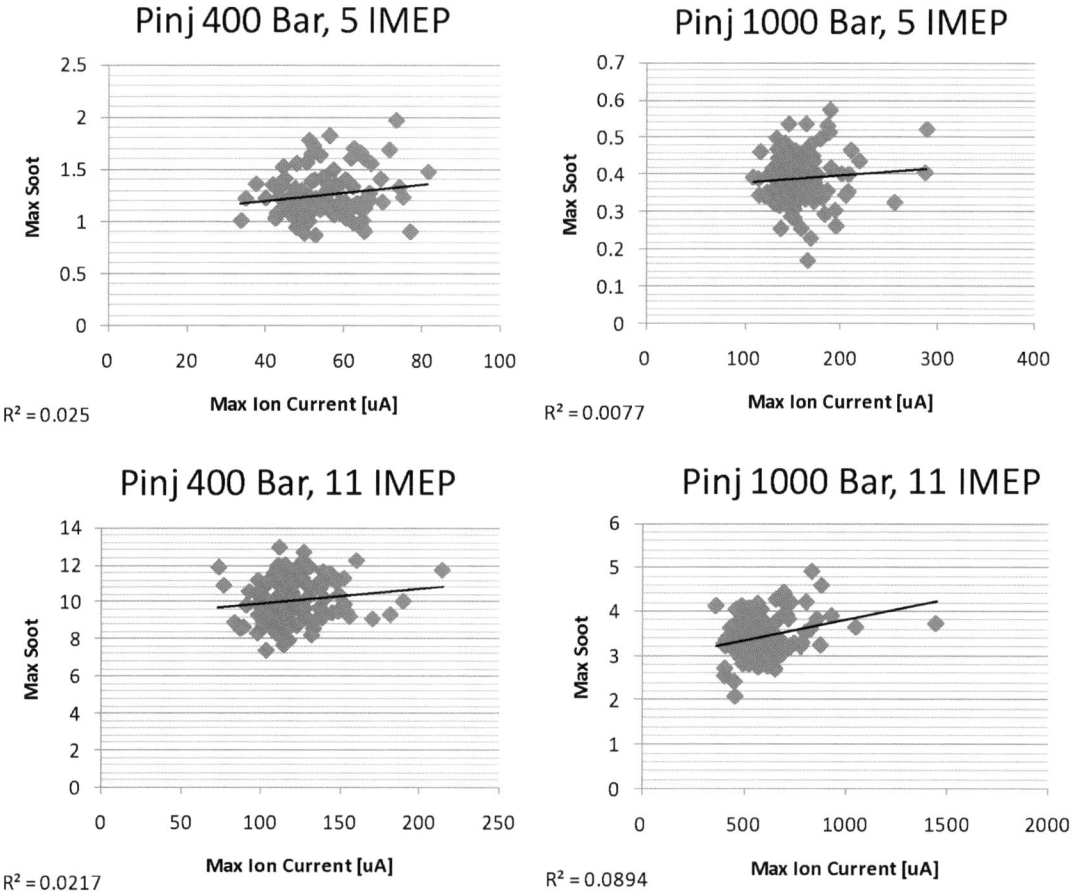

Figure 5.4   Correlation between maximum ion current and soot at low and high injection pressure and engine load on a cyclic basis.

5.4 ENGINE TRANSIENT OPERATION WITH OPEN AND CLOSED ECU

A test was conducted on the heavy duty John Deere engine using an open ECU under transient engine speeds, loads and intake pressures. The speed was varied from 1150 RPM to 2000 RPM while the torque fluctuated between 70 Nm and 220 Nm. The figure shows traces for varying torques, manifold absolute pressures and speeds. The lowest figure shows two traces for the soot percentage. A new technique was used to predict soot from the ion current signal. Based on this investigation, a close match is achieved between the measured and predicted soot as shown in FIG 5.5.

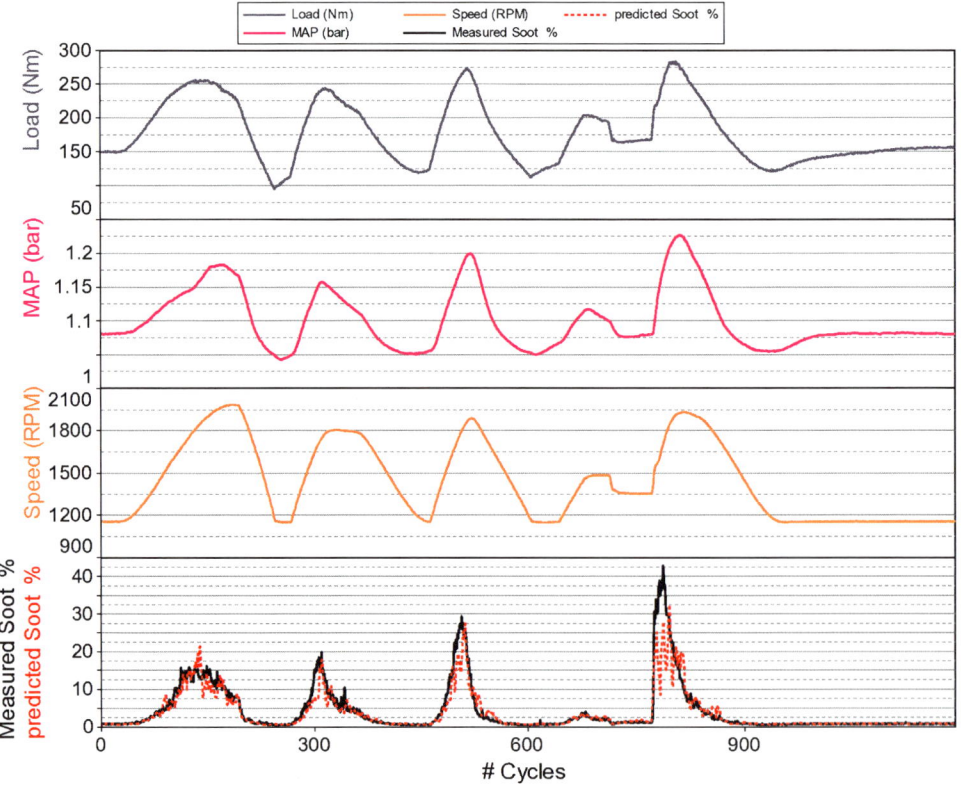

Figure 5.5   Traces for instantaneous load, MAP, speed, and percentage of soot measured by the opacity meter and computed from the ion current signal.

A second test was carried out to compare between predicted and measured soot using the original manufacturer closed ECU. In this test, the following engine operating parameters vary: load, intake pressure, injection pressure, and injection timing. Only engine speed was kept constant.

The original manufacturer ECU used for this test was calibrated by the manufacturer to produce soot emissions within the EPA standards. The test was developed to see if the predicted soot using the new technique is sensitive enough to capture the very low soot levels emitted. The engine speed was kept constant at 1800 RPM, load (IMEP) varied between 12 and 18 bar, injection pressure varied between 950 and 1150 bar, and intake pressure (MAP) varied between 2.4 and 2.8 bar. The results in FIG 5.6 showed a good match between the measured and predicted soot ranging between 0.05% and 0.6%.

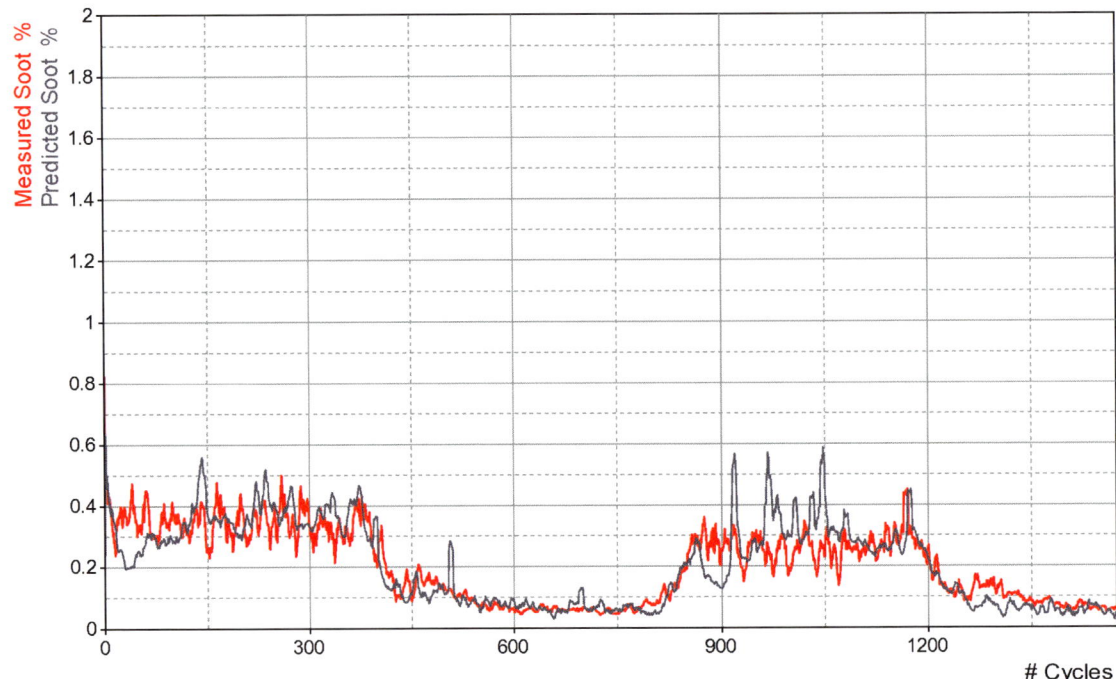

Figure 5.6   Correlation between measured and predicted soot using a closed ECU.

5.5 ADVANTAGES OF THE SOOT PREDICTION TECHNIQUE

The main advantage of the new soot prediction technique is to offer a cost effective method to determine soot concentration inside the combustion chamber using the ion current signal. The technique also provides a fast cycle-by-cycle soot computation to cope with the engine transient operation and high speeds. The feedback signal sent to the engine ECU can be used to adjust different engine operating parameters in order to produce less soot to comply with the EPA stringent emissions standard. A list of advantages the soot prediction technique is provided as follows:

1. Cost effective as the sensor involved is the ion sensor.

2. Fast response soot measuring technique, as it depends on electron speed.

3. Measures soot inside the combustion chamber.

4. Measure soot on a cycle-by-cycle basis.

5. Measure soot in each engine cylinder.

6. No modifications required to the engine block.

7. Onboard tool for soot determination.

8. Compact design.

5.6 <u>CONCLUSIONS</u>

- Experiments were designed to study the effect of soot on the ion current signal at different engine operating conditions.

- The correlation between ion current signal amplitude and soot measured in the exhaust on a cycle by cycle basis turned out to be fairly weak. Therefore, another correlating technique was developed.

- A new soot prediction technique was established based on the ion current signal. The technique was tested on a diesel engine at different operating conditions during steady and transient operation.

CHAPTER 6   IN-CYLINDER NO AND ION CURRENT CORRELATIONS

6.1 <u>CHAPTER OVERVIEW</u>

Ion current sources in diesel engines are not well identified [79]. The purpose of this chapter is to study the effect of in-Cylinder NO concentration on the ion current at various engine operating conditions. The chapter introduces a novel technique of ion current measurement which involves the patented in-cylinder NO sampling-probe / ion-current sensor [80] discussed in Chapter 4.

Ion current is a local signal and it is important to sample the cylinder gases for NO measurements at the same exact location of the ion current probe inside the combustion chamber. This allows us to determine experimentally the contribution of NO in the ion current. Furthermore, this arrangement will help in determining if one of the ion current peaks is due to only NO thermal ionization, NO chemical ionization, or a mix of both and what is their ratio. Also to find out if the ion current is from a different source. The experiments were conducted on a heavy duty multi cylinder John Deere diesel engine under different engine parameters such as injection pressure and load.

66

6.2 EXPERIMENTAL SETUP

The experiments throughout this chapter were conducted on a multi-cylinder heavy duty John Deere diesel engine. The engine is equipped with a common rail injection system and Variable Geometry Turbocharger (VGT). The engine was controlled using a full accessed open ECU, where engine parameters such as injection timing, injection pressure, intake pressure, and engine speed were controlled.

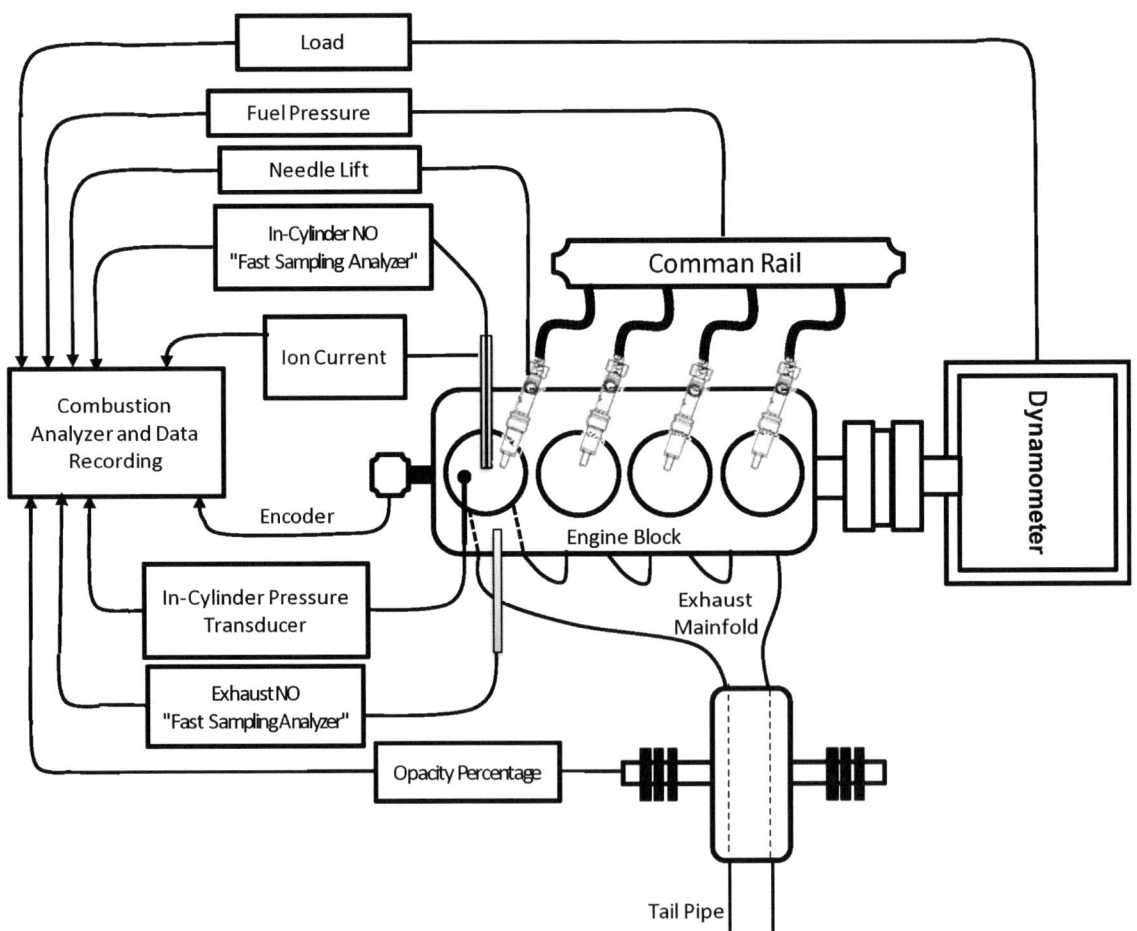

Figure 6.1   Experimental layout used for NO and ion current measurements.

A Kistler pressure transducer was fitted in the head of cylinder 1 to measure the cylinder gas pressure. An optical encoder was installed on the front end to acquire the crank angle position with a resolution of 0.25 CAD. Engine load, needle lift, intake pressure, and intake temperature were also acquired. The engine out emissions such as soot and NO were measured. An opacity meter measures the soot content of the exhaust gases. A fast response sampling analyzer (CLD 500) was connected to the exhaust pipe for NO measurements. In addition, the CLD 500 analyzer was connected to the in-cylinder NO sampling-probe / ion-current sensor fitted in cylinder 1 to measure in-cylinder NO content as well as ion current. Figure 6.1 shows engine layout and instrumentation.

A test matrix was designed to study the contribution of NO measured inside the combustion chamber in the ion current signal. The test matrix is shown in Table 6.1. The engine test was performed at a low injection pressure of 400 bar as well as high injection pressure of 1000 bar. A sweep of load of 5, 7, 9, and 11 bar IMEP was conducted at each injection pressure. Speed is maintained constant at 1300 RPM. In addition, the Variable Geometry Turbocharger (VGT) was used to maintain the same intake pressure during tests. Figure 6.2 shows cylinder pressure traces, needle lift signal, calculated rate of heat release, and cylinder temperature at 400 bar (top graph) and 1000 bar (bottom graph) injection pressure .

TABLE 6.1

Injection Pressure	400 bar		1000 bar	
Engine Load	SOI	Fuel Flow (g/min)	SOI	Fuel Flow (g/min)
5 bar	8 bTDC	82.3	8 bTDC	75
7 bar	8 bTDC	117	8 bTDC	108
9 bar	8 bTDC	163	8 bTDC	131
11 bar	8 bTDC	197	8 bTDC	173

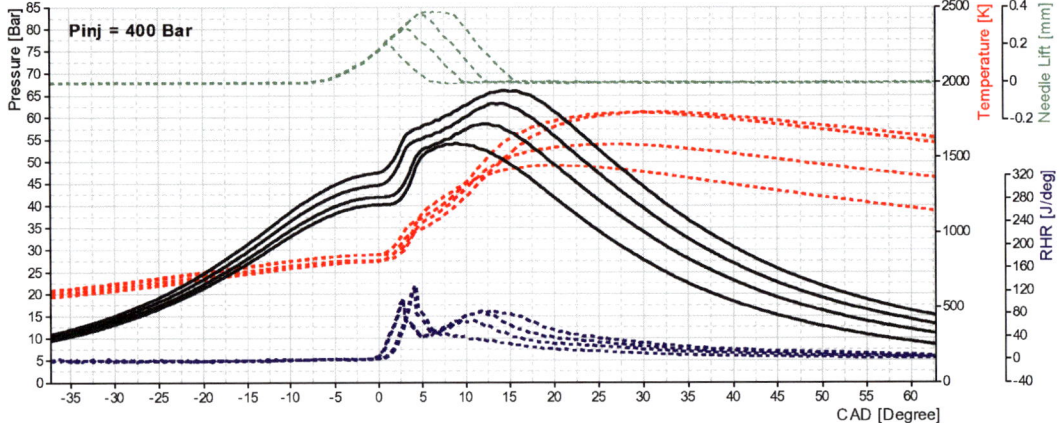

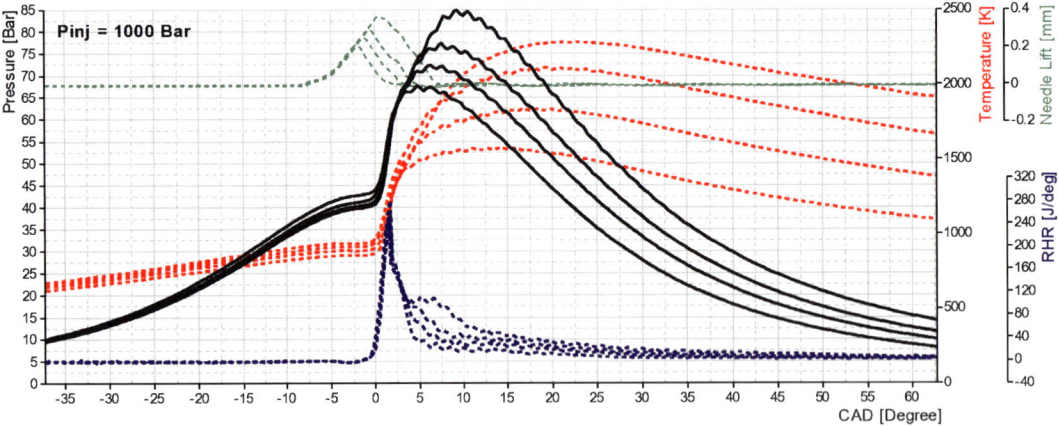

Figure 6.2 Recorded cylinder pressure traces (Black) and needle lift signal (Green), and calculated rate of heat release (Blue) and cylinder temperature (Red) for a sweep of engine load at 2 injection pressures, 400 bar (Top figure) and 1000 bar (Bottom figure).

6.3 <u>AVERAGED ION CURRENT SIGNAL ANALYSIS</u>

Figure 6.3 and FIG 6.4 represent the experimental results of the John-Deere diesel engine at different loads and injection pressures. Each trace is an average of 100 cycles. Ion current is recorded against In-Cylinder NO. Different engine loads at 5, 7, 9, and 11 bar IMEP are shown at two injection pressures, 400 and 1000 bar. The engine speed is 1300 RPM. The relationship between measured ion current and in-cylinder NO reflected the same trend at low and high injection pressures. As engine load increased, both ion current and NO increased until 9 bar IMEP. Nevertheless, at higher loads, 11 bar, NO decreased but the ion current peak kept increasing.

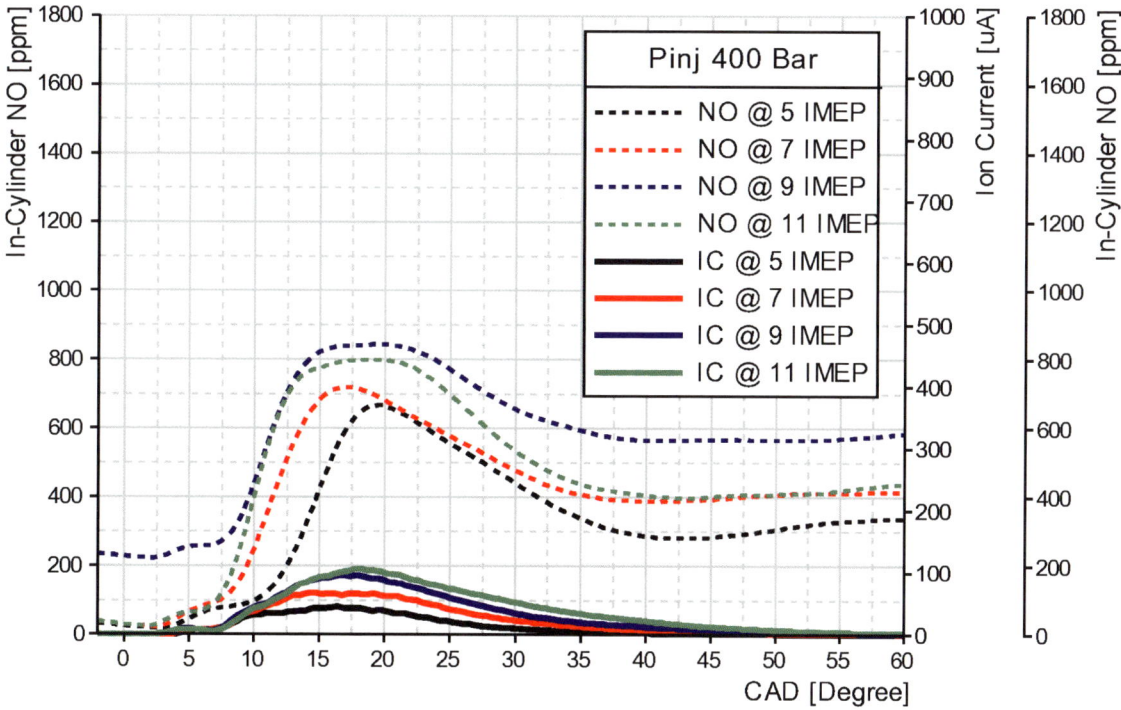

Figure 6.3   Average traces of 100 cycles for In-Cylinder NO concentration and ion current signal (IC) at 400 bar injection pressure with varying load.

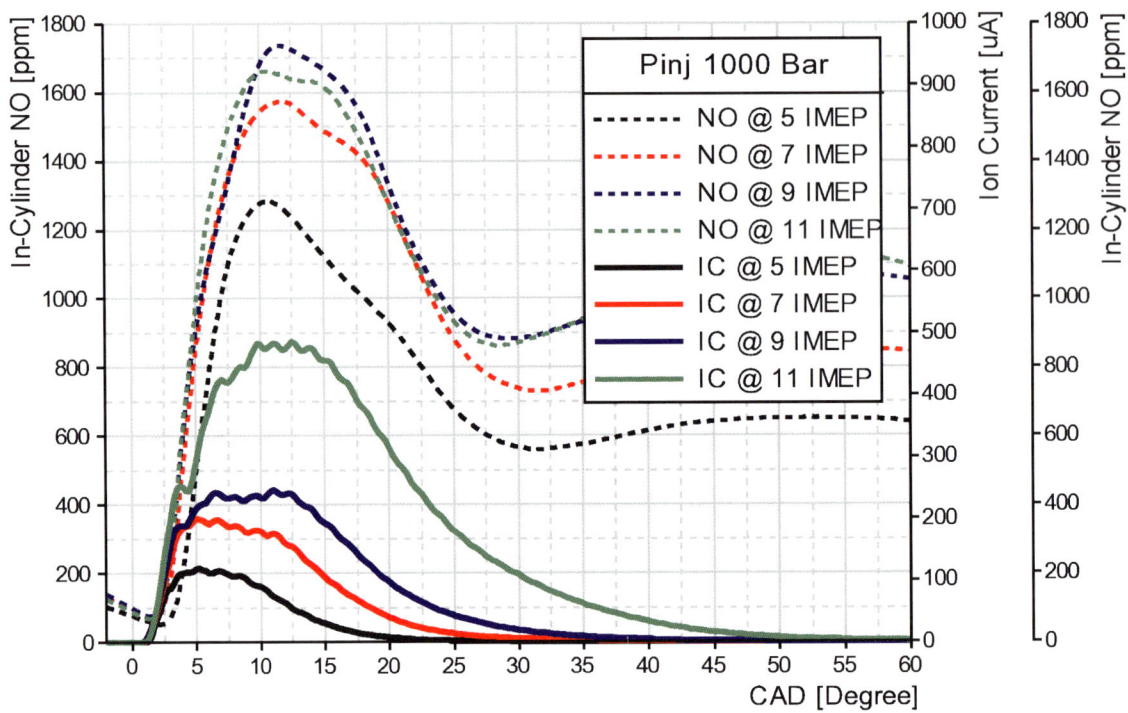

Figure 6.4   Average traces of 100 cycles for In-Cylinder NO mole fraction and ion current signal (IC) at 1000 bar injection pressure with varying load.

6.4 <u>THERMAL ION CURRENT CALCULATION</u>

In SI engines, NO by itself is responsible for 95% of the available free electrons in the post flame gases, which is responsible for the second ion-current peak   [5, 6, 81-84]. Saitzkoff and Reinmann developed a mathematical model to predict the ion current second peak amplitude [5, 6, 24, 85] based on NO thermal ionization. Chemi-ionization was not accounted for in this equation. The model was derived from the statistical physics Saha's equation and used calculated cylinder temperature and assumed In-Cylinder NO concentration as input [6, 24, 86].

$$I = \left(\frac{U}{d}\right)(\pi DL)\left(\sqrt{\frac{2\left(\frac{2\pi m_e kT}{h^2}\right)^{\frac{3}{2}}\exp\left[-\frac{E_i}{kT}\right]}{n_{tot}}}\sqrt{\varphi_s}\right)\left(\frac{e^2}{m_e\sigma\sqrt{\frac{8kT}{\pi m_e}}}\right) \qquad (6.1)$$

Where, (I) is the ion current, (U) is the applied voltage, (d) is the ion sensor gap, (D) and (L) are the ion sensor diameter and length, (m_e) is the electron mass, (k) is Boltzmann constant, (σ) is the electron cross-section, (T) is temperature, (h) is Plank's constant, (E_i) is the activation energy, (Φ_s) is the mole fraction of the species s, (e) is the electron charge, (n_{tot}) is the total number density of the gas.

Thermal ion current calculations are conducted based on NO measured inside the cylinder and the calculated cylinder temperature by using equation 6.1. However, the accuracy of the calculations is affected by a major fact. The temperature used as an input to the mathematical model is the cylinder average temperature calculated from the pressure trace. This will result in lower calculated values of thermal ion current than in reality. The reason behind this is the fact that in diesel combustion local temperatures can reach much higher values than average ones due to the heterogeneity of the charge.

A sensitivity analysis is carried out to show the effect of NO and temperature on thermal ion current calculations. Figure 6.5 reflects these effects. Figure (6.5a) shows the thermal ion current calculated based on varying NO from 100 ppm to 10,000 ppm while maintaining the temperature constant at 2300 K. The thermal ion current is

$100*10^{-6}$ A at 1000 ppm. It reached a maximum of $273*10^{-6}$ A at 10,000 ppm. It is clear that even if NO is increased 10 times, the thermal ion current does not vary that much.

Turning to figure (6.5b), NO is kept constant at 1000 ppm while temperature changed from 2300 K to 4200 K. Thermal ion current increased from $100*10^{-6}$ A at 2300 K to 10 A at 4200 K. This analysis shows that thermal ion current is extremely sensitive to temperature rather than NO. Raising the temperature 2 times led to significant jump in the thermal ion current. On the other hand, the ion current did not change much by raising NO 10 times. This conclusion leads to the fact that small distortions and errors occurring while sampling the In-Cylinder NO will have minimal effect on the thermal ion current calculations.

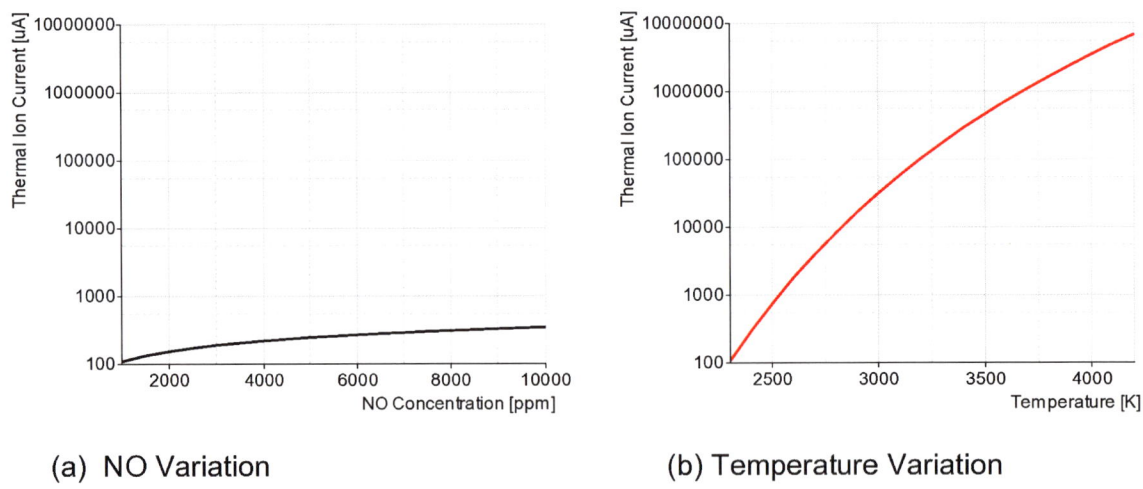

(a) NO Variation   (b) Temperature Variation

Figure 6.5   Sensitivity analysis comparing thermal ion current sensitivity based on NO and temperature variation.

Figure 6.6 shows 100 cycles averaged experimental results from the engine as well as calculated thermal ion current (TIC) at 400 bar and 1000 bar injection pressure with varying load. Calculated average cylinder temperatures and NO concentrations measured inside the combustion cylinder are also plotted in FIG 6.6.

At 1000 bar injection pressure (FIG 6.6a), thermal ion current (TIC) picked up as the cylinder temperature increased with load. The highest thermal ion current occurred at 11 bar IMEP (Highest Load). Thermal ion current drops to zero with temperatures below 2000 K which is the case at 5 bar and 7 bar IMEP. Turing to 400 bar injection pressure (FIG 6.6b), thermal ion current is almost negligible at all engine loads. This is caused by the fairly low cylinder temperature even at the highest load.

As a conclusion, thermal ion current (TIC) increases as the injection pressure goes up. In addition, TIC becomes more significant at higher loads due to increase in cylinder gas temperature. This investigation indicates that there is a temperature window at which NO starts to ionize thermally. Furthermore, TIC follows the temperature trend rather than NO.

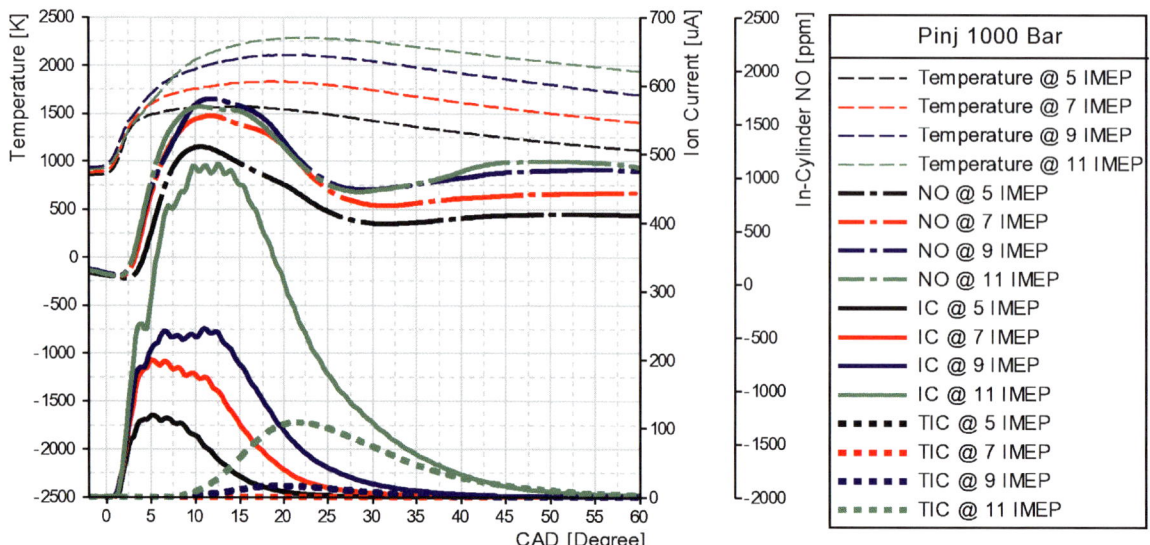

(a) 1000 bar injection pressure

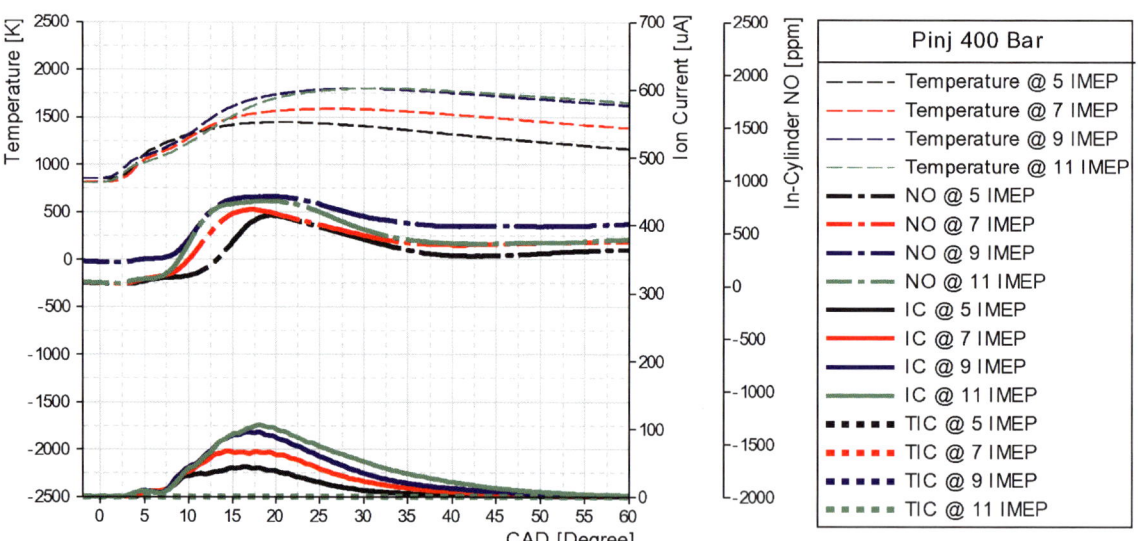

(b) 400 bar injection pressure

Figure 6.6   Averaged traces for cylinder temperatures, In-Cylinder NO, and calculated

thermal ion current (TIC) at different loads.

One final check on thermal ion current calculation is to verify the possibility of using the in-exhaust NO signal instead of in-cylinder NO as a simplified method to calculate TIC. Table 6.2 represents a comparison between in-cylinder NO and In-exhaust NO mole fractions measured simultaneously in this investigation at various engine operating conditions. Based on the results reported in this table, it is clear that NO mole fractions measured inside the combustion chamber are at the most 2 times that measured in the exhaust port.

TABLE 6.2

Injection Pressure	1000 bar		400 bar	
Engine Load	NO In-Cylinder	NO In-Exhaust	NO In-Cylinder	NO In-Exhaust
5 bar	1283 [ppm]	872 [ppm]	652 [ppm]	372 [ppm]
9 bar	1722 [ppm]	1062 [ppm]	839 [ppm]	483 [ppm]
11 bar	1655 [ppm]	957 [ppm]	800 [ppm]	422 [ppm]

Figure 6.7 shows a comparison between thermal ion current (TIC 1) calculated based on in-cylinder NO signal and (TIC 2) calculated based on in-exhaust NO signal (measured in exhaust port) at 1000 bar injection pressure and 11 bar IMEP. A fairly good agreement is shown between TIC 1 and TIC 2. Thus, this test shows the ability to use NO concentration obtained from exhaust measurements in thermal ion current calculations.

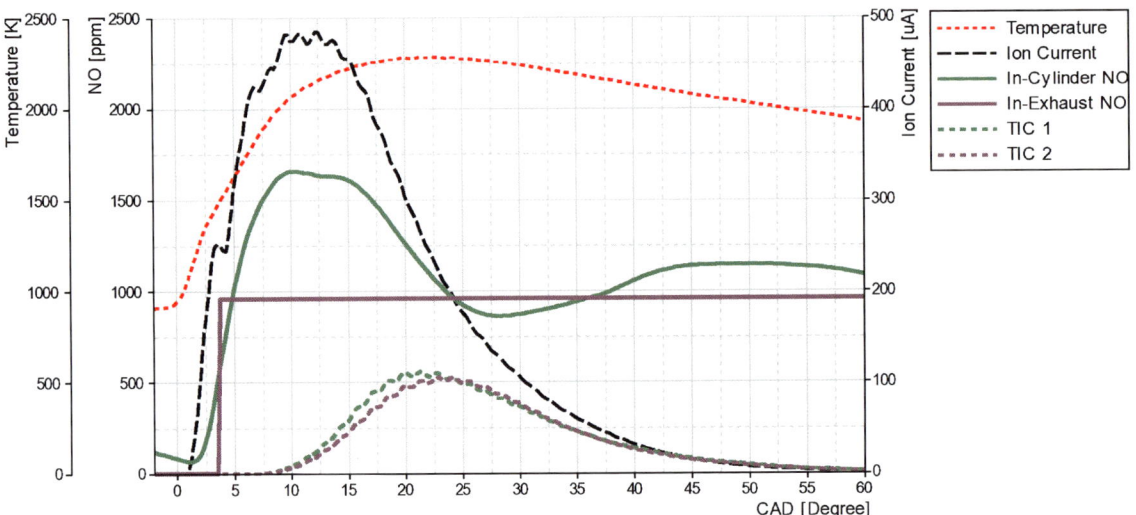

Figure 6.7   Comparison between thermal ion current calculated from In-Cylinder NO signal and In-Exhaust NO.

6.5 CYCLE-BY-CYCLE ION CURRENT SIGNAL ANALYSIS

This section reflects cycle-by-cycle variations in the ion current signal recorded throughout the experiments under the conditions given in Table 6.1. Thermal ion current (TIC) based on NO thermal ionization is calculated from both in-cylinder NO and in-exhaust NO signals at different engine loads and injection pressures.

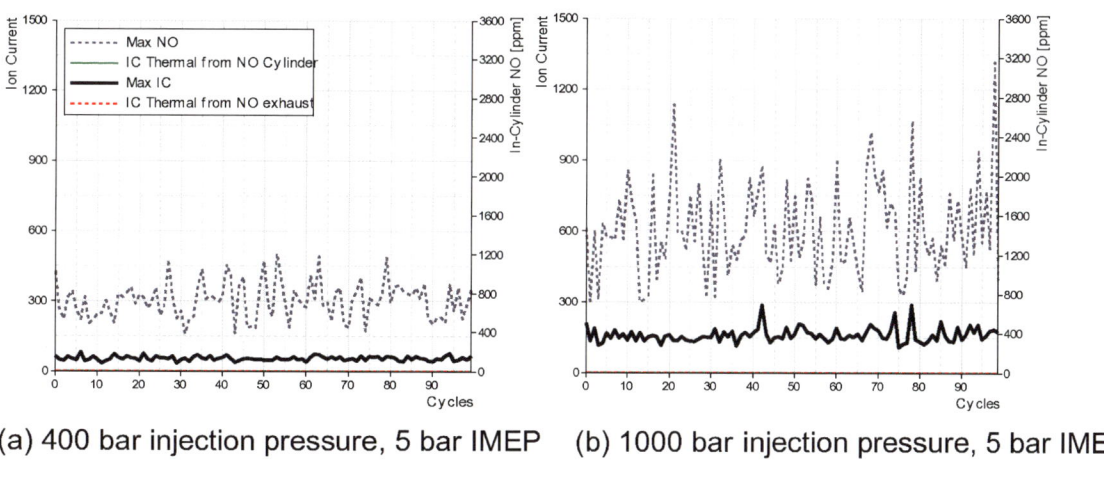

(a) 400 bar injection pressure, 5 bar IMEP   (b) 1000 bar injection pressure, 5 bar IMEP

(c) Pinj at 400 bar, 11 bar IMEP   (d) Pinj at 1000 bar, 11 bar IMEP

Figure 6.8   A set of graphs including ion current, NO, and calculated thermal ion current peaks on a cyclic basis.

As a first step to study the correlation between the ion current signal and in-cylinder NO, FIG 6.8 shows ion current peak (Max IC) versus in-cylinder NO peak (Max NO) in each cycle. The figure consists of 4 graphs representing lowest (5 bar IMEP) and highest (11 bar IMEP) loads at each injection pressure. In FIG (6.8a), Pinj 400 bar and 5 bar IMEP, maximum ion current varies between 40 and 80 uA, while peak in-cylinder NO fluctuates from 400 to 1200 ppm. Both NO and IC peaks increased slightly at same injection pressure and higher load as shown in FIG (6.8c). Turning to high injection pressure, FIG (6.8b) and FIG (6.8d) represent Pinj 1000 bar at 5 bar and 11 bar IMEP. A dramatic increase in NO concentration is shown in both figures. Furthermore, ion current recorded its highest values at high injection pressure and high load.

Calculated thermal ion current along all engine cycles is insignificant in all cases except one. In FIG (6.8d), high engine load and high fuel injection pressure, IC Thermal fluctuates around 100 uA, close to 15% of the total ion current signal. Thermal ion current presence is one factor that explains ion current signal (max IC) high values shown in FIG (6.8d). Furthermore, the comparison between thermal ion current calculations based on in-cylinder and in-exhaust NO represents a close match as shown in the figure.

The data in FIG 6.8 was re-plotted as shown in FIG 6.9 to come up with the correlation factor in each case. The correlation factor corresponding to FIG (6.8a), (6.8b), (6.8c), (6.8d) is 28.6%, 40.4%, 9.9%, and 2% respectively. These percentages show that the correlation between NO peaks and IC peaks deteriorates as load is increased at all injection pressures. The reason behind this is the increase in soot levels at high engine loads which becomes part of the total ion current amplitude as discussed in Chapter 5.

On the other hand, at low load, a better correlation is achieved as injection pressure is increased. This is caused by the increase in NO concentration inside the combustion chamber at higher injection pressures as the combustion is enhanced due to better fuel atomization. The worst correlation between NO and IC peaks takes place at Pinj 1000 bar and 11 bar IMEP. Besides NO, other factors affect the ion current signal in this case such as high soot amounts and high thermal ion current values.

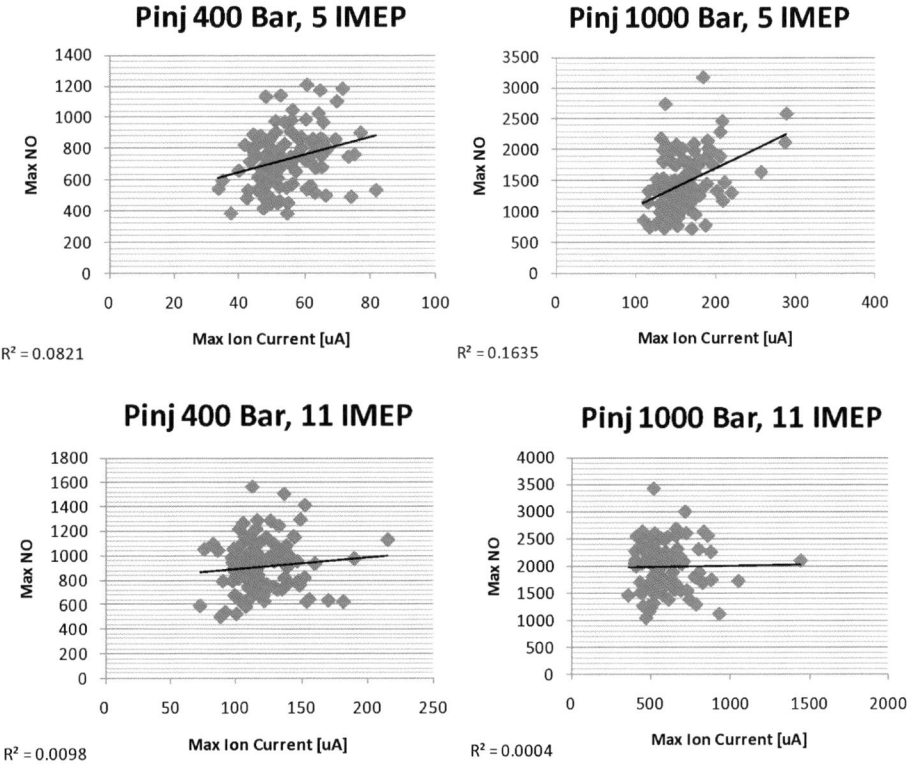

Figure 6.9   Correlation between maximum ion current and In-Cylinder NO at low and high injection pressure and engine load on a cyclic basis.

Another analysis is conducted to show the relationship between in-Cylinder NO peaks, Cylinder temperature peaks and thermal ion current peaks on a cycle-by-cycle basis at 1000 bar injection pressure and 11 bar IMEP.   Figure (6.10a) shows a correlation factor of 29.88% between thermal ion current and in-cylinder NO. However, the correlation factor is 90.87% between cylinder temperature and thermal ion current as shown in FIG (6.10b). These results are in good agreement with the sensitivity analysis performed in section 6.4.

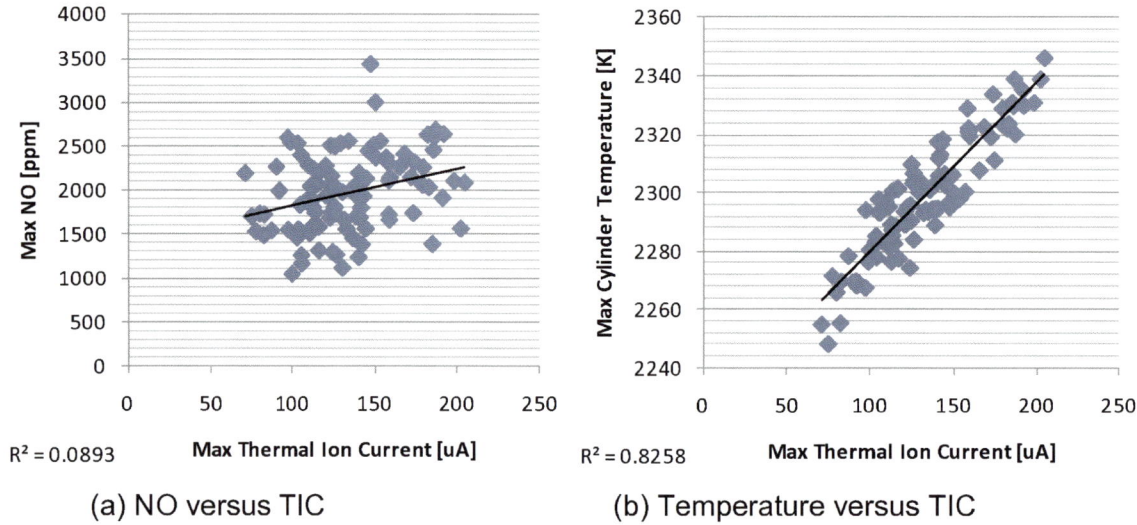

$R^2 = 0.0893$ **Max Thermal Ion Current [uA]** $R^2 = 0.8258$ **Max Thermal Ion Current [uA]**

(a) NO versus TIC   (b) Temperature versus TIC

Figure 6.10   Correlation between thermal ion current (TIC) and In-Cylinder NO and cylinder temperature on a cyclic basis.

6.6 NO EFFECT ON ION CURRENT SIGNAL PROFILE

This section reveals new findings correlating in-cylinder NO to ion current signal shape on a cyclic basis at different engine operating conditions. Various cycles are shown at different engine loads and injection pressures. This novel analysis is fairly accurate as In-Cylinder NO and ion current were measured simultaneously. Moreover, this type of investigation requires ion current to be measured at the same exact location where NO is sampled inside the cylinder. This is important because ion probes are considered local sensors. Figure 6.11 (Pinj 1000, 11 bar IMEP) shows traces for the cylinder gas pressure, calculated mass average gas temperature, NOx in the sample, and the measured ion current (IC) corresponding to that specific cycle. Thermal ion current (TIC) is also calculated. Despite the poor correlation (2%) between NO and IC maximum amplitudes, depicting the same number of peaks in the signals of the NO and ion current traces is a major finding that has been observed in all the cycles recorded. Figure 6.11 shows a sample of cycles with one to three peaks in the two signals.

Nevertheless, Figure 6.11 differentiates between the contribution of NO ionized chemically (NO^+ Chemi) and NO ionized thermally (NO^+ Thermal) to the ion current signal. Thermal NO^+ is calculated from in-cylinder NO, average cylinder temperature and equation 6.1. First, NO^+ Thermal is responsible of thermal ion current (TIC) production as discussed before. TIC starts at 12 CAD aTDC and peaks at 22 CAD aTDC in most cycles as it follows cylinder temperature. Second, NO^+ Chemi follows In-Cylinder NO profile. It reflects the same number of peaks in the ion current signal. In addition, a phase shift is clearly observed between Chemi and Thermal NO^+ peaks.

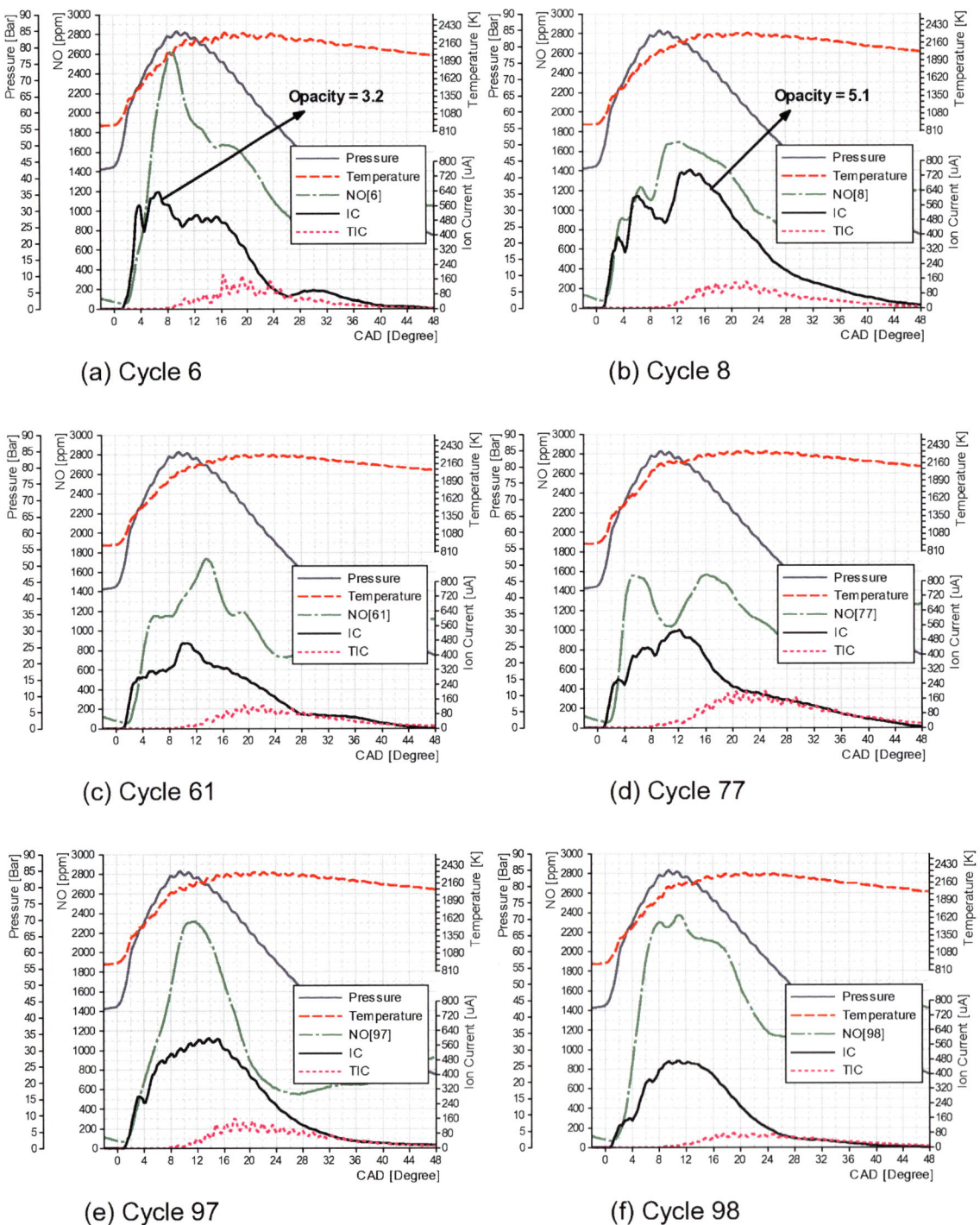

(a) Cycle 6

(b) Cycle 8

(c) Cycle 61

(d) Cycle 77

(e) Cycle 97

(f) Cycle 98

Figure 6.11   Cycle-by-Cycle Variation at 1000 bar injection pressure and 11 bar IMEP.

Figure 6.12 represents different cycles at 1000 bar injection pressure and 5 bar IMEP. The correlation between ion current and NO profiles can still be observed. Nevertheless, NO^+ Thermal is insignificant in this case as cylinder temperature is low. During this engine operating condition, soot production is minimal as discussed in Chapter 5. This reduces soot interference on the ion current signal dramatically. The major contributor to the ion current signal at high injection pressure and low load is mainly NO^+ Chemi.

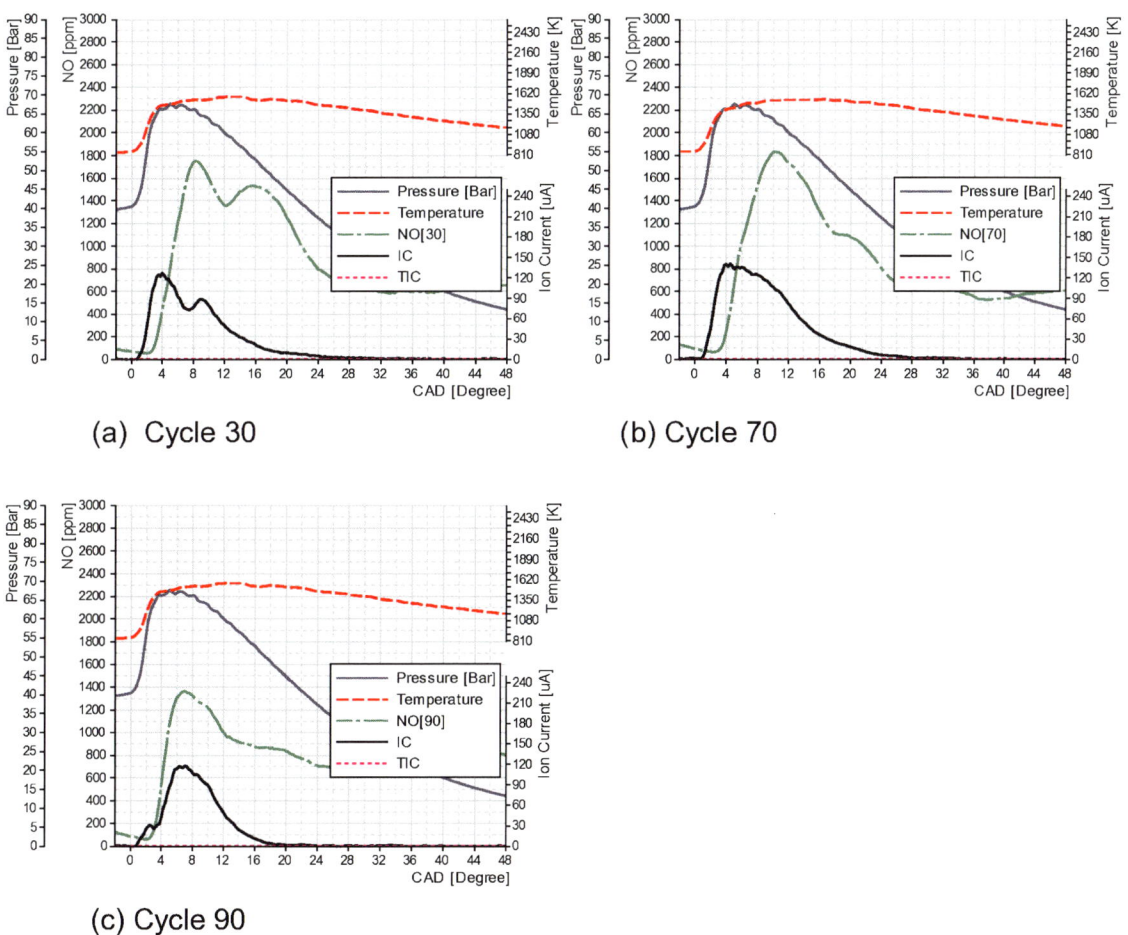

   (a) Cycle 30   (b) Cycle 70

   (c) Cycle 90

Figure 6.12   Cycle-by-Cycle Variation at 1000 bar injection pressure and 5 bar IMEP.

84

Figures 6.13 represents engine cycles at 400 bar injection pressure and high load. Figure 6.14 represents cycles at 400 bar injection pressure and low load. For both cases, the correlation between NO and ion current profiles is lost. Chemi NO$^+$ effect is minimal at this fairly low injection pressure. Thermal NO$^+$ is insignificant.

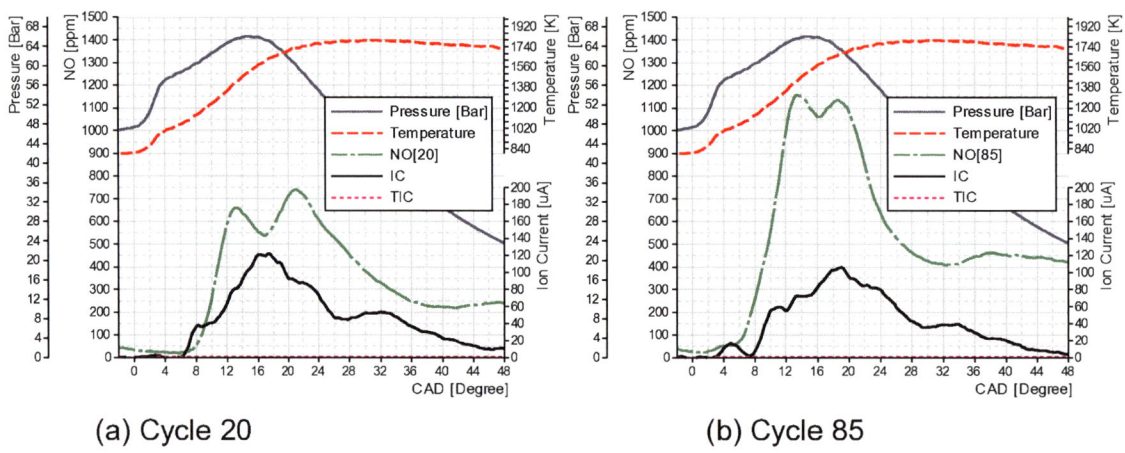

(a) Cycle 20　　　　　　　　　　　　　(b) Cycle 85

Figure 6.13　Cycle-by-Cycle Variation at 400 bar injection pressure and 11 bar IMEP.

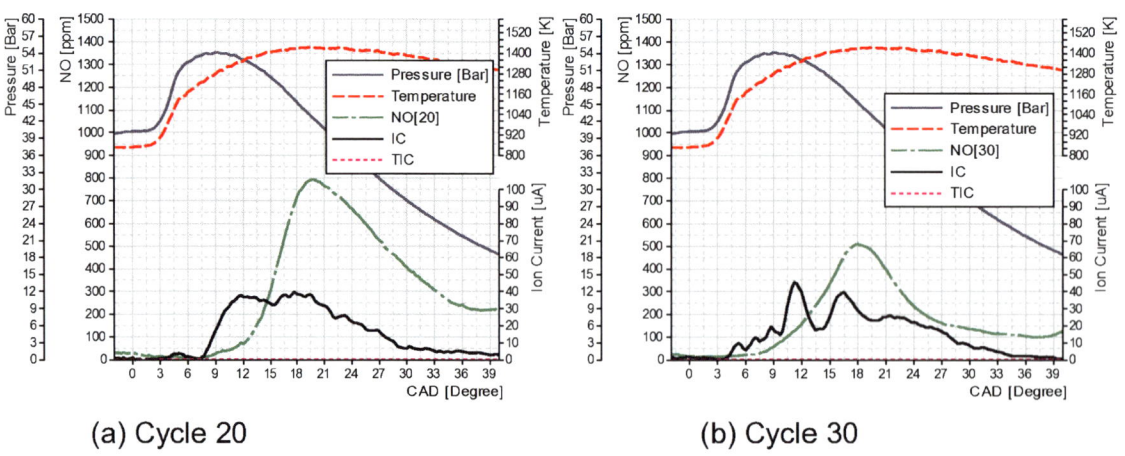

(a) Cycle 20　　　　　　　　　　　　　(b) Cycle 30

Figure 6.14　Cycle-by-Cycle Variation at 400 bar injection pressure and 5 bar IMEP.

6.7 CONCLUSIONS

- Experiments were designed to study the effect of NO concentration on the ion current signal. An in-cylinder NO gas sampling probe was converted to work as an ion current sensor in order to study that effect.

- Ion current based on NO thermal ionization was calculated using cylinder temperature and in-cylinder NO concentration. The thermal ion current was compared to that calculated based on NO concentration measured in the exhaust port. Both methods gave the relatively close results as the thermal ion current is sensitive to temperature rather than NO concentration.

- Ion current was studied on a cycle by cycle basis. A strong correlation was found between the number of peaks in the ion current signal and that in the in-cylinder NO signal at high injection pressures. However, at low injection pressures, this correlation does not hold.

- A differentiation can be made between NO ionized due to thermal processes and that ionized due to chemical processes. The peak ion current due to NO thermal ionization is well correlated with the peak cylinder temperature. The peak ion current due to NO chemi ionization is related to the peaks occurring in the in-cylinder NO signal.

CHAPTER 7   MECHANISM OF IONIZATION IN DIESEL ENGINES

7.1 <u>CHAPTER OVERVIEW</u>

Ionization in combustion engines produces a signal indicative of in-cylinder conditions that can be used for the feedback electronic control of the engine, to meet the production goals in performance, fuel economy and emissions. Most of the research has been conducted on spark ignition engines where the ionization mechanisms are well defined. A limited number of investigations have been conducted on ionization in diesel engines because of its complex combustion process.

This chapter presents a mechanism the author has developed for ionization in diesel engines considering the heterogeneity of the charge and the resulting variations in the combustion products. The mechanism accounts for the wide variability in the equivalence ratio from very lean to near stoichiometric, rich and soot producing mixtures. The mechanism is introduced in a diesel cycle simulation code to determine the contribution of different species in the ionization process. In addition, the relative contribution of the chemi-ionization of NO as compared to the well known its thermal ionization is determined.

7.2 PROPOSED IONIZATION MECHANISM IN DIESEL COMBUSTION

The proposed mechanism is based on a set of elementary reactions chosen from a variety of sources and references. The set of reactions along with the reaction rate coefficients describe the rate of formation and dissociation of various ionic structures as well as neutral species. Furthermore, the proposed mechanism considers chemi-ionization, ion-molecule charge transfer, and charge recombination reactions as well as reactions representing thermal ionization of certain species. The proposed mechanism will be referred to as "diesel ion formation mechanism", (DIF). DIF consists mainly of 62 ionic reactions including 26 ionic species.

7.2.1 Ionization Reactions

Early ions production is based on the chemi-ionization exothermic initiation reaction (R1) which is widely accepted as the source of ions in hydrocarbon flames [57, 58, 129- 135]. CHO^+ is assumed to be the source of all ions found in the charge during combustion. However, it is not the dominant ion and its lifetime is extremely short because it is rapidly consumed in charge transfer ion-molecule reactions [24, 58, 92-95].

$$CH + O \leftrightarrow CHO^+ + e \qquad (R1)$$

Various nitrogenous ions such as NO^+, NH_3^+, and NH_4^+ are also added to the pool of ions. Furthermore, detailed chemi-ionization reactions of NO are included through charge transfer reactions with CHO^+, H_3O^+, O_2^+, N_2^+, N^+, CO^+, CO_2^+, $C_2H_2^+$, and $C_2H_4^+$.

DIF also considers thermal ionization process of 4 species (NO, CH_3, CHO, C_3H_3) represented in the form of a chemical reaction (R2 to R5). The reactions backward third-order charge recombination specific reaction rate constant is known [104-106]. Equilibrium constant is calculated using Saha's equation and thus the forward specific reaction rate constant is determined for all four reactions. The reaction activation energy in this case is equal to the ionization potential of the thermally ionized species.

$NO + M \leftrightarrow M + NO^+ + e$ (R2)

$CH_3 + M \leftrightarrow M + CH_3^+ + e$ (R3)

$CHO + M \leftrightarrow M + CHO^+ + e$ (R4)

$C_3H_3 + M \leftrightarrow M + C_3H_3^+ + e$ (R5)

NO, CH_3, CHO, C_3H_3 are chosen from a set of 21 species considered for thermal ionization processes. The author investigated the contribution of many species in the thermal ion current at different equivalence ratios as shown in FIG 7.1. C_3H_3 and NO are the major contributors to the thermal ion current at equivalence ratio 0.4 as shown in FIG (7.1a). Figure (7.1b) reflects NO domination (99%) at equivalence ratio 1.0. However, in FIG (7.1c), NO share drops to 35% while CH_3, CHO, C_3H_3 are considered at higher equivalence ratios, 2.2. Nevertheless, FIG (7.1d) shows that on the fairly rich side, at $\Phi = 3.0$, C_3H_3 ionization dominates the thermal ion current.

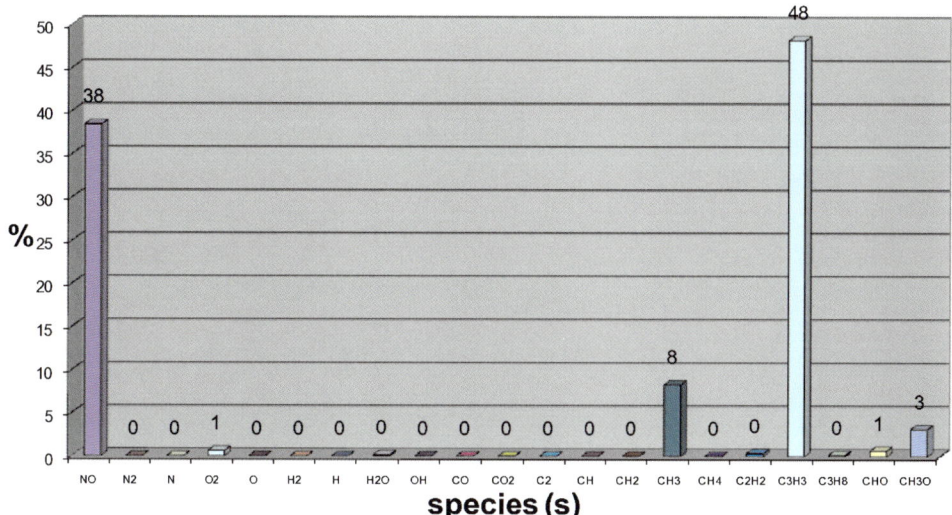

(a) Equivalence Ratio = 0.4

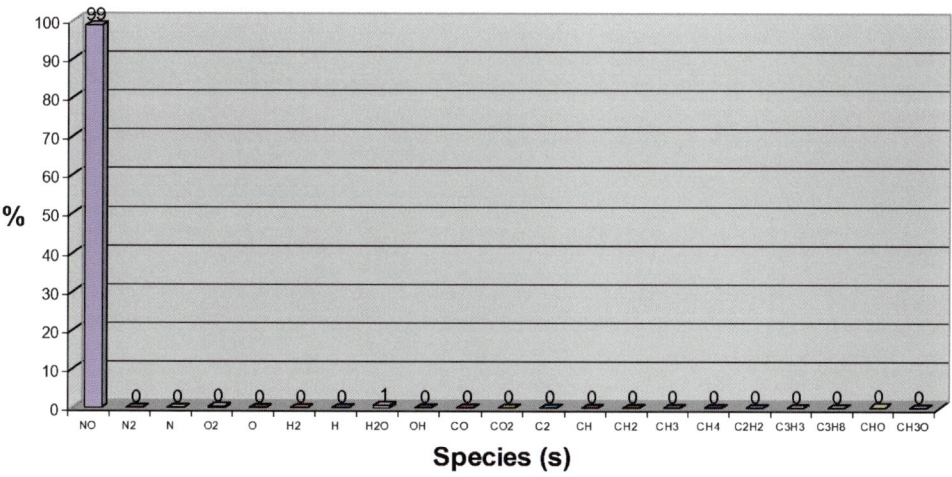

(b) Equivalence Ratio = 1.0

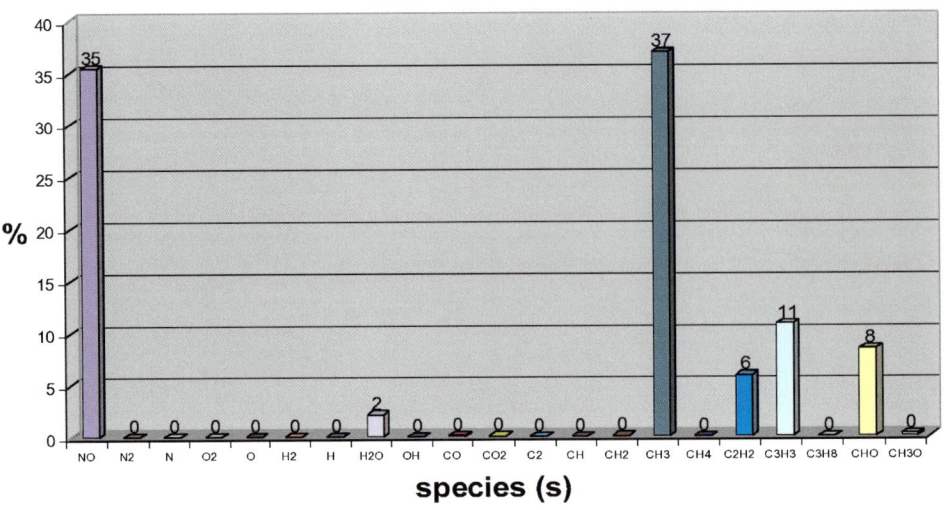

(c) Equivalence Ratio = 2.2

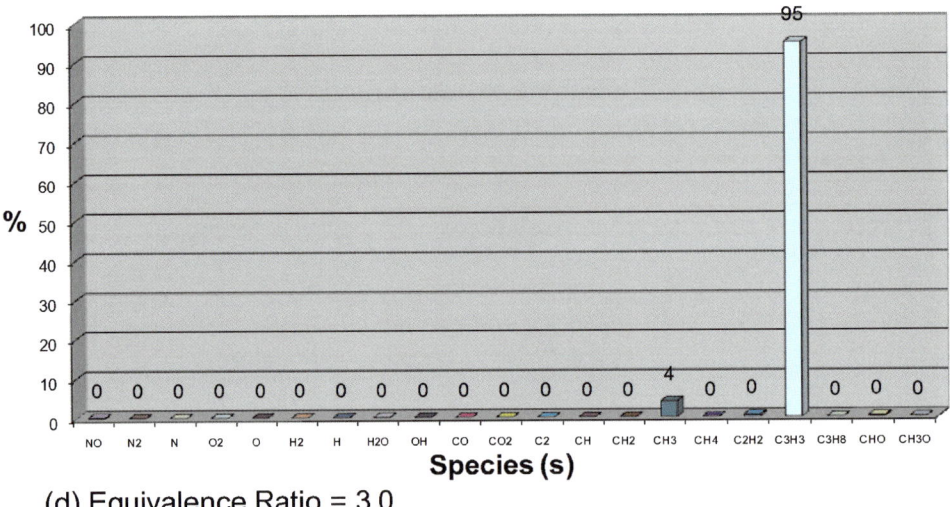

(d) Equivalence Ratio = 3.0

Figure 7.1   Representation of the contributors to thermal ionization at equivalence ratio

0.4 1.0 (Top Graph), and equivalence ratio 2.2 (Bottom Graph) 3.0.

7.2.2 <u>Thermochemical Data</u>

Thermochemical database required for neutral and ionized species used in the DIF model is based on three properties, heat of formation, entropy, and specific heat. The values of these properties are highly affected by gas temperature; therefore JANAF tables [107, 108], NASA [109], and other references [57, 59, 110- 115] were used to gather these data.

Obtaining thermochemical data for neutral species is straightforward. On the other hand, gathering ions database at various temperatures is a difficult task to accomplish. Limited references provided information about ions thermochemical properties such as enthalpy of formation. In cases where some data are absent, estimations were made. That is based on the assumption that specific heats of ions are equal to the value for the corresponding neutral species as the contribution of a single electron to the specific heat of a molecule is expected to be very small [57, 59]. The same concept is applied to calculate ions entropies whenever the database is unavailable.

7.2.3 <u>Diesel Fuel Surrogate and Combustion Mechanism</u>

As commonly recognized, real fuels are complex mixtures of thousands of hydrocarbon compounds including linear and branched paraffins, naphthenes, olefins and aromatics. It is generally agreed that their behavior can be effectively reproduced by simpler fuel surrogates containing a limited number of components [116, 117, 121- 124]. Hence, n-heptane, a primary reference fuel for octane rating in internal combustion engines, is chosen in this study to model combustion and transport processes taking place in diesel engines. N-heptane has a cetane number of approximately 56, which is slightly higher than the cetane number of conventional diesel fuel.

The main fuel oxidation mechanism used for this study is the Lawrence Livermore National Laboratory (LLNL) n-heptane detailed mechanism, version 3 [118]. It contains 2540 reversible elementary reactions among 557 intermediate species. This detailed mechanism performs well at both low and high temperatures and over a broad range of pressures common to diesel engines. The mechanism is based on the previously developed and very successful mechanism of Curran et al. 1998 [119]. DIF is used along with the LLNL n-heptane mechanism to compute the concentrations of ions and intermediate combustion species in the diesel engine.

7.2.4 Diesel Cycle Simulation Code

CHEMKIN-PRO, a powerful system to solve complex chemical kinetics problems, is selected for this study. This zero-dimensional software, generally favored for creating new mechanisms, is used to build the new DIF model. The internal combustion engine reactor (ICR) is picked to simulate the auto-ignition conditions in the diesel engine [119]. Variations in the volume of the ICR are calculated via equations provided by Heywood that describe the volume as a function of time, based on engine parameters, including compression ratio, crank radius, connecting rod length, speed of revolution of the crank ,shaft and the clearance or displaced volume [137].

This closed system reactor is a single-zone model assuming homogeneous charge contained in an adiabatic cylinder. Although these assumptions do not totally represent a real diesel engine with heterogeneous mixtures, this research is carried out based on the fact that a diesel engine burns pockets of flammable mixtures at different equivalence ratios affecting the ion current signal. Each of these pockets is studied separately within the internal combustion engine reactor [92].

7.3 CYCLE RESOLVED ANALYSIS OF IONIZATION IN DIESEL COMBUSTION

The DIF is applied in a diesel cycle simulation to investigate in some details the proposed ionization mechanisms. The heavy duty John Deere diesel engine configurations are used in this investigation. The engine has a 106 mm cylinder bore, 127 mm stroke and 17.5 compression ratio. The start of fuel injection is at 4 crank angle degrees before TDC. N-heptane, in gaseous state, is assumed to be injected and form a homogeneous charge which burns according to the LLNL mechanism. The effect of EGR (Exhaust Gas Recirculation) is not included in this study. The simulation covers a wide range of mixtures, starting from a fairly lean to close to stoichiometric, slightly rich, and extremely rich mixtures. The equivalence ratio varied from 0.2 to 4.0. The following figures show the results obtained from Chemkin simulation using the DIF model associated with LLNL n-heptane combustion mechanism.

Figure 7.2 shows the peak values of the cylinder gas pressure, temperature, and the concentration of the electrons at each equivalence ratio. The electrons concentration is fairly low at equivalence ratio of 0.2, increases sharply and reaches a peak at an equivalence ratio slightly higher than the stoichiometric. This is followed by a sharp drop at higher equivalence ratios and reached a minimum at Φ = 1.75. A small peak in the electrons concentration is observed at Φ = 2, the critical equivalence ratio for soot formation. The reason behind this will be discussed in details.

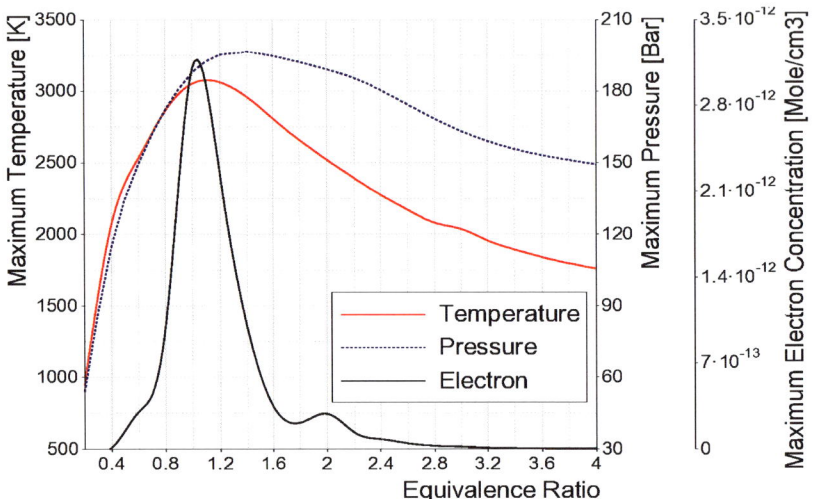

Figure 7.2   Peak values of cylinder gas pressure, temperature, and electron concentration at different equivalence ratios.

7.3.1 Effect of equivalence ratio on total ion concentration

Figure (7.3a), (7.3b), and (7.3c) show the cylinder gas pressure and electrons concentration from the start of injection at 4° bTDC to 40° aTDC at different equivalence ratios. Pressure traces are plotted in dotted lines while electrons concentrations are plotted in solid lines. Notice the variation in the vertical scale among the figures, made to show the shapes at different equivalence ratios.

Figure (7.3a) is for Φ = 0.6 to Φ = 1.2. The pressure traces reflect the long ignition delay for the very lean charge caused by low temperature combustion regime

(LTCR). The ignition delay is reduced as the equivalence ratio increases. As shown in the traces, electrons start to increase sharply with the increase in pressure in the high temperature combustion regime (HTCR). Figure (7.3b) and (7.3c) reflect the decrease of ion current as the cylinder pressure drops as equivalence ratio increases. The start of the ion current can be considered as an indication of the start of combustion at all equivalence ratios. Reaction (R1) shows clearly that CH radical is the initiator of the chemi-ionization process, as shown in FIG 7.4. The high concentration of the CH radical in diesel combustion has been reported from observations made on an optically accessible engine [41, 42].

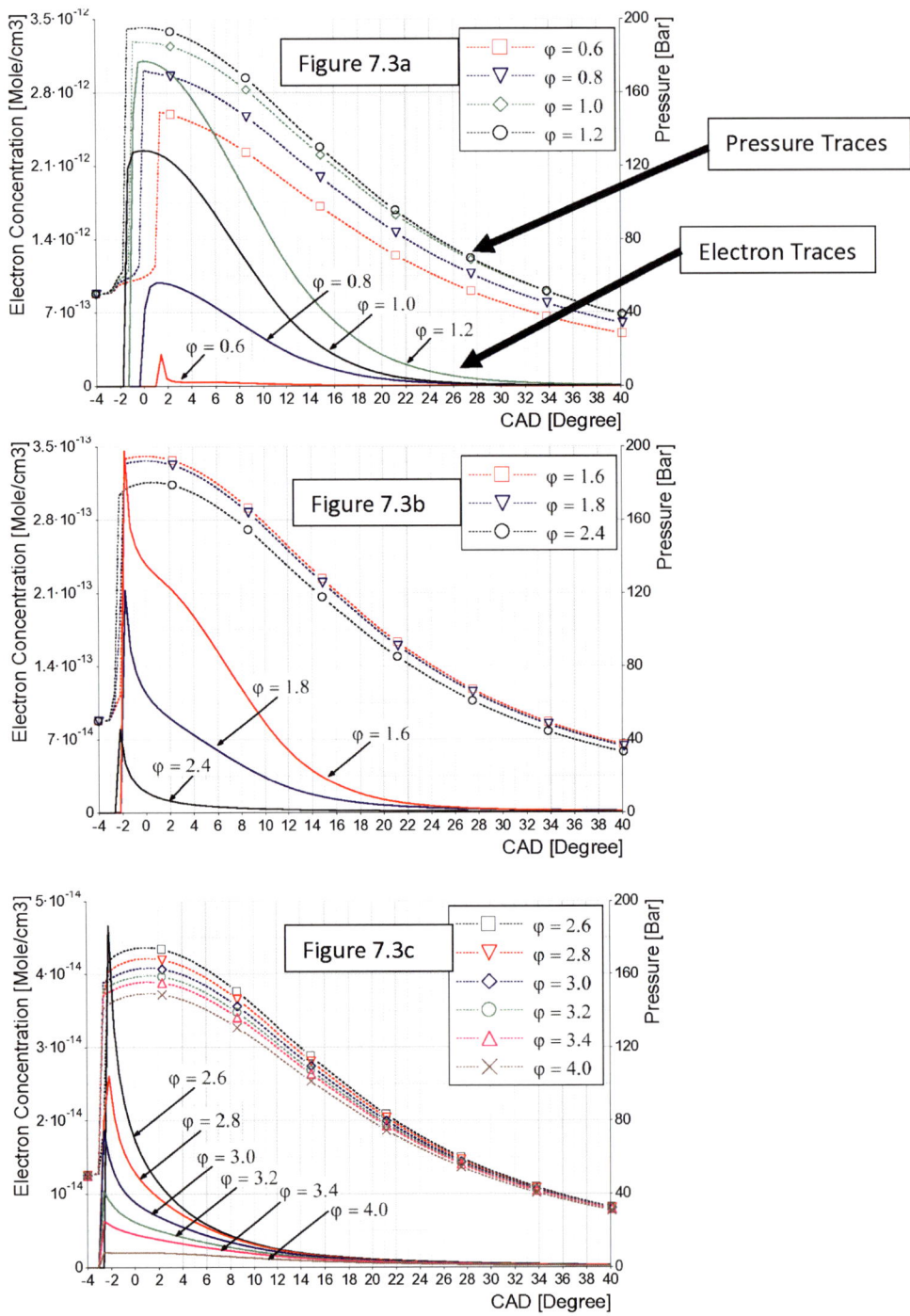

Figure 7.3   Cylinder gas pressure and electron concentration traces at different equivalence ratios.

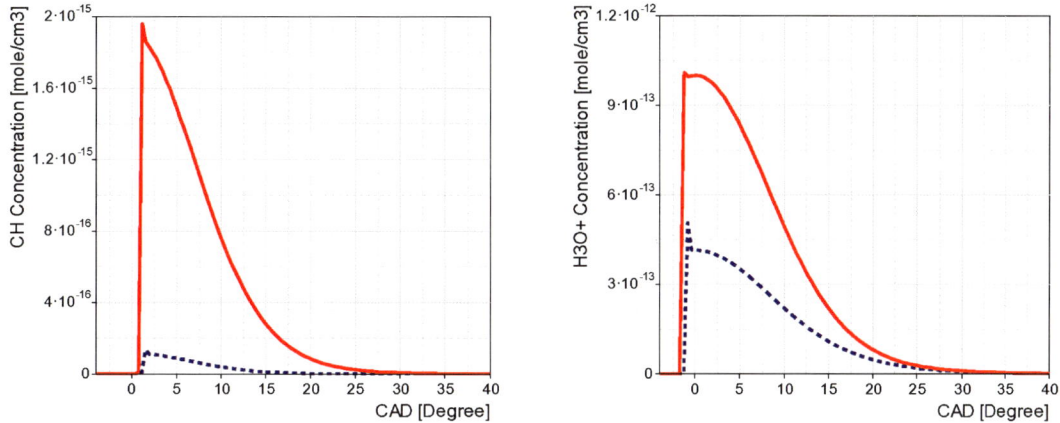

Figure 7.4   Comparison between CH (Left Graph) and H3O+ (Right Graph) concentrations at 2 different equivalence ratios 0.8 and 1.2.

7.3.2 Effect of Equivalence Ratio on Ionized Species Development

Figure 7.5 is a more detailed set of figures showing the contribution of different species in the diesel ionization process at equivalence ratios that vary from 0.4 to 4.0. From all the 29 ionic species shown in section 7.2, only a few showed dominance. The dominant ions have longer lifetime because they are rapidly formed by consuming other ions through charge transfer reactions. Moreover, they have low rate of consumption. H3O+ dominates the ion current at fairly lean mixtures such as equivalence ratio 0.4. NO+ is the major ionic species around stoichiometry.

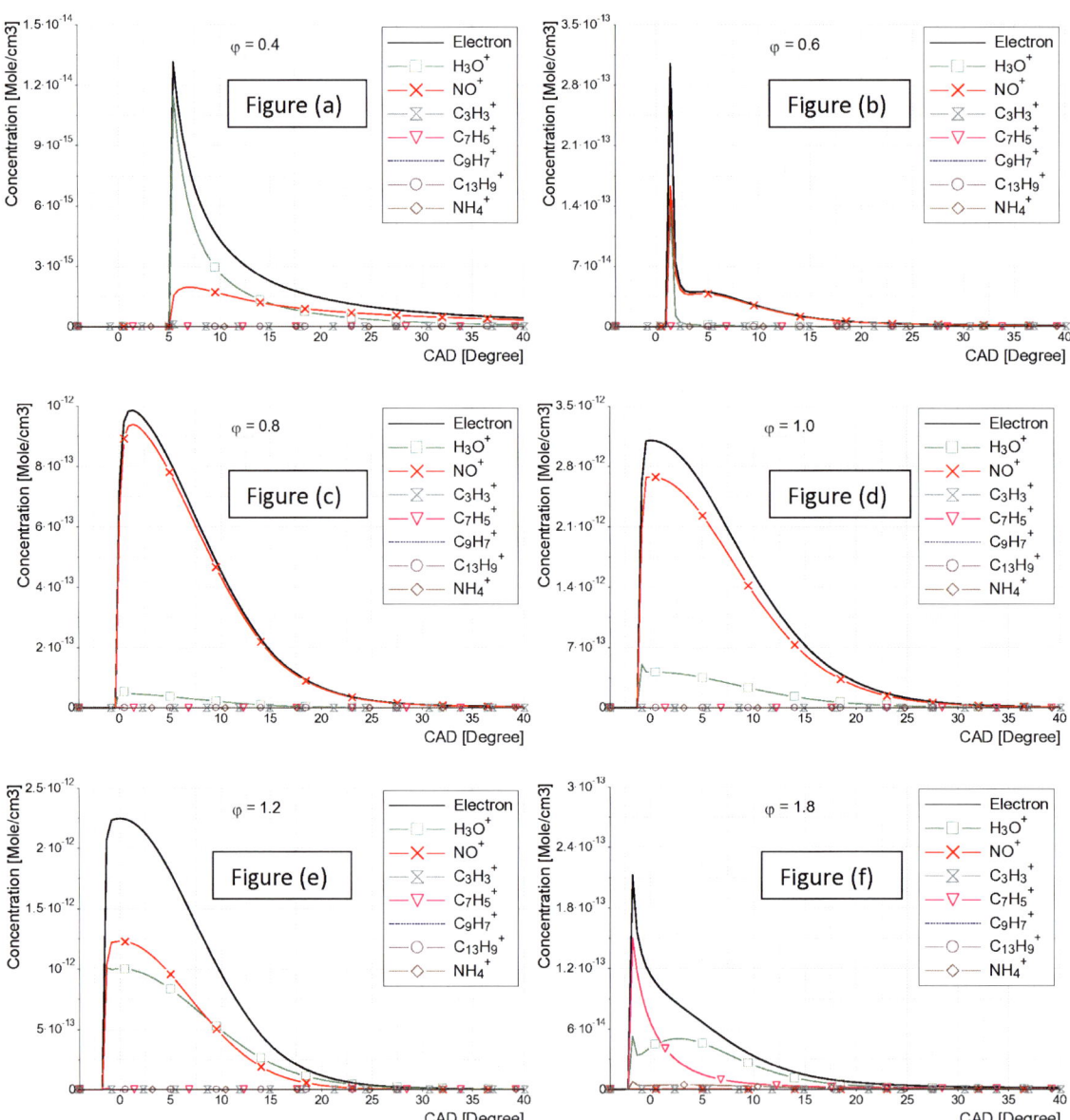

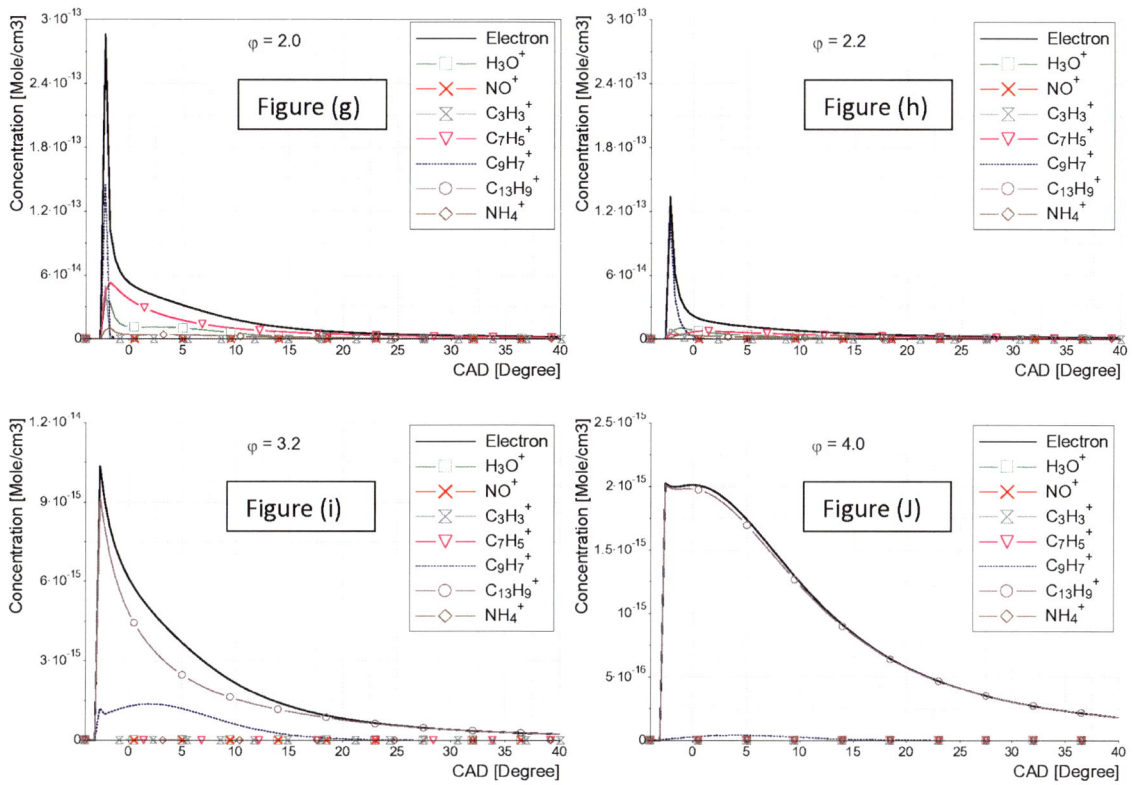

Figure 7.5   Detailed development of ionized species at different equivalence ratios.

7.3.3 Effect of equivalence ratio on major ionized species concentration

Figure 7.6 shows the peak values of the major ionic species at different equivalence ratio. Figure (7.6a) is a representation of the dominant ionized species concentration under a linear vertical scale while FIG (7.6b) reflects the same information under a logarithmic vertical scale. NO^+ production is the highest around equivalence ratio of unity. H_3O^+ reached maxima at Φ equals 1.2.

It is interesting to notice the sharp rise in the concentration of C9H7+ ion as the equivalence ratios increases from 1.8 to 2.0. This explains the source of the bump in the electron concentration observed previously in FIG 7.2 at Φ equals 2.0. From FIG (7.6b), it is clear that this bump resulted from the sharp rise in $C_9H_7^+$ concentration from 10^{-18} at Φ = 1.8 to 10^{-13} at Φ = 2.0. Meanwhile, H_3O^+ and $C_7H_5^+$ concentrations did not change significantly.

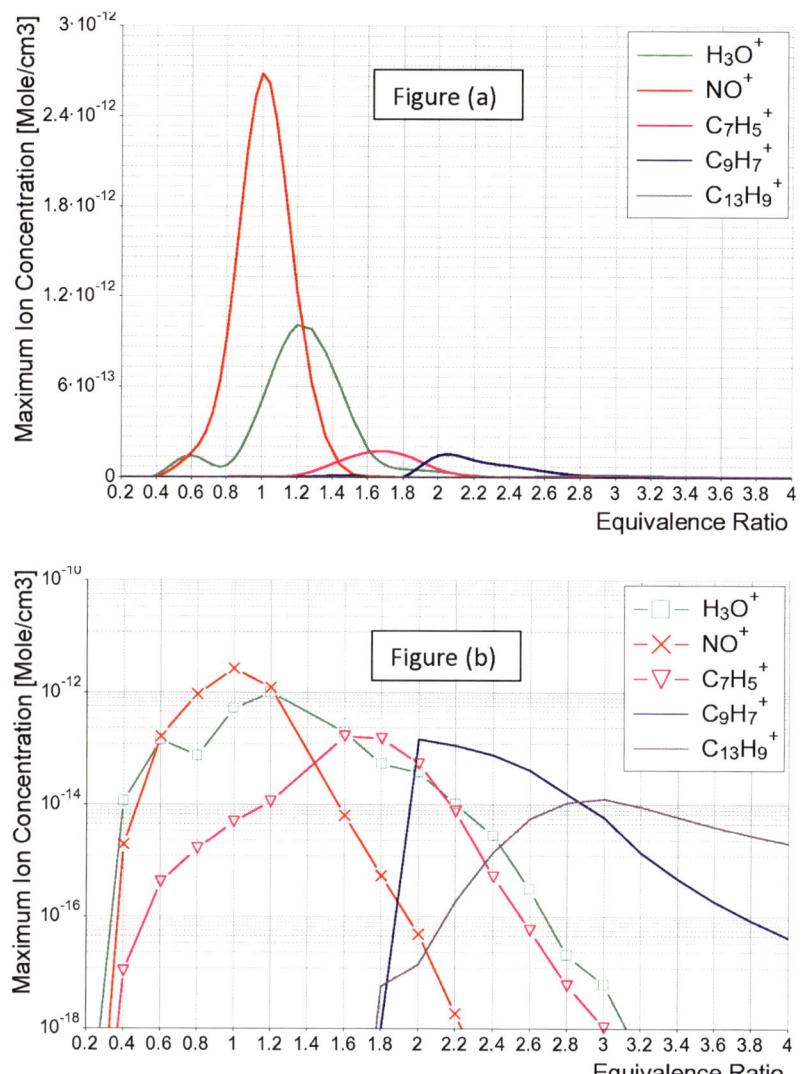

Figure 7.6   Peak values of the major ionic species at different equivalence ratio.

Figure 7.7 is a 3-D plot of the concentration of the major ions from the TDC to 40°aTDC, for different equivalence ratios. H_3O^+, $C_7H_5^+$, $C_9H_7^+$, and $C_{13}H_9^+$ are shown. This figure reflects an increase in the concentration of the heavier hydrocarbon ions at the higher equivalence ratios as discussed before.

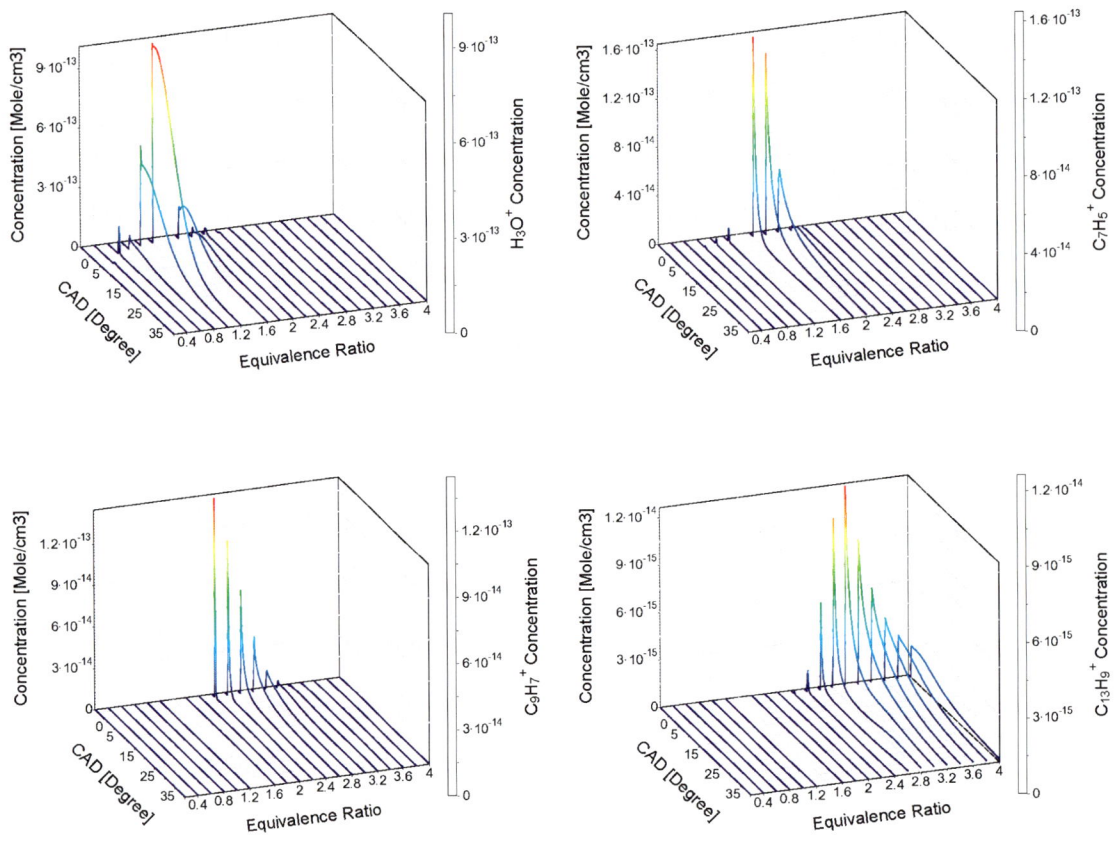

Figure 7.7   3-D Plots for the concentration of the hydrocarbon ions from the TDC to 40°

aTDC at different equivalence ratios.

7.4 RELATIVE CONTRIBUTION OF CHEMICAL AND THERMAL IONIZATION

Ion current in diesel engines is mainly formed by chemi-ionization and thermal-ionization processes. The ratio between these two processes has not been studied before. This section investigates the effect of equivalence ratio on chemi and thermal ionization. Figures (7.8a) and (7.8b) show the peak values and the percentage of the peak values of the thermal and chemical processes in the total concentration of ions at different equivalence ratios.

Figure 7.8 indicates the following:

(a) Chemi-ionization is dominant at equivalence ratios up to 0.6 and higher than 1.0.

(b) The contribution of thermal ionization starts at Φ equal 0.8 and diminishes at Φ equals 1.6.

(c) Thermal-ionization is dominant from equivalence ratio 0.8 to 1.0.

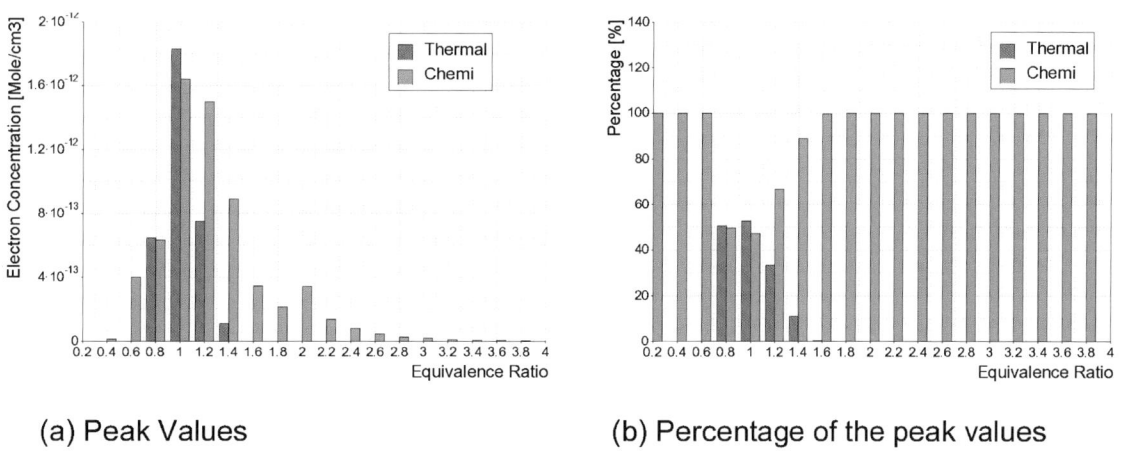

(a) Peak Values   (b) Percentage of the peak values

Figure 7.8   Contribution of the thermal and chemical processes in the total electrons concentration at different equivalence ratios.

Figure 7.9 is a set of detailed graphs showing the breakdown of the total ion current signal at different equivalence ratios. Total electron traces, as well as thermal and chemical traces are plotted against crank angle degree. The contributions of thermal and chemical signals vary as the equivalence ratio changes. For Φ equal 0.8 and 1.0, thermal signal surpasses the chemical signal. However, at 1.2 and 1.6 equivalence ratios, chemical signal is more significant.

Regarding the thermal traces composition, NO^+ is the dominant ion resulted from the thermal ionization process at equivalence ratios 0.8 to 1.6. CHO^+ starts to contribute to the thermal signal along with NO^+ at Φ equals 1.8. On the other hand, CH_3^+, CHO^+, and $C_3H_3^+$ are the major contributor to the thermal ionization signal from Φ equal 2.0 to 4.0. However, the thermal ionization process is insignificant at equivalence ratios higher than 1.6 due to much lower cylinder temperatures.

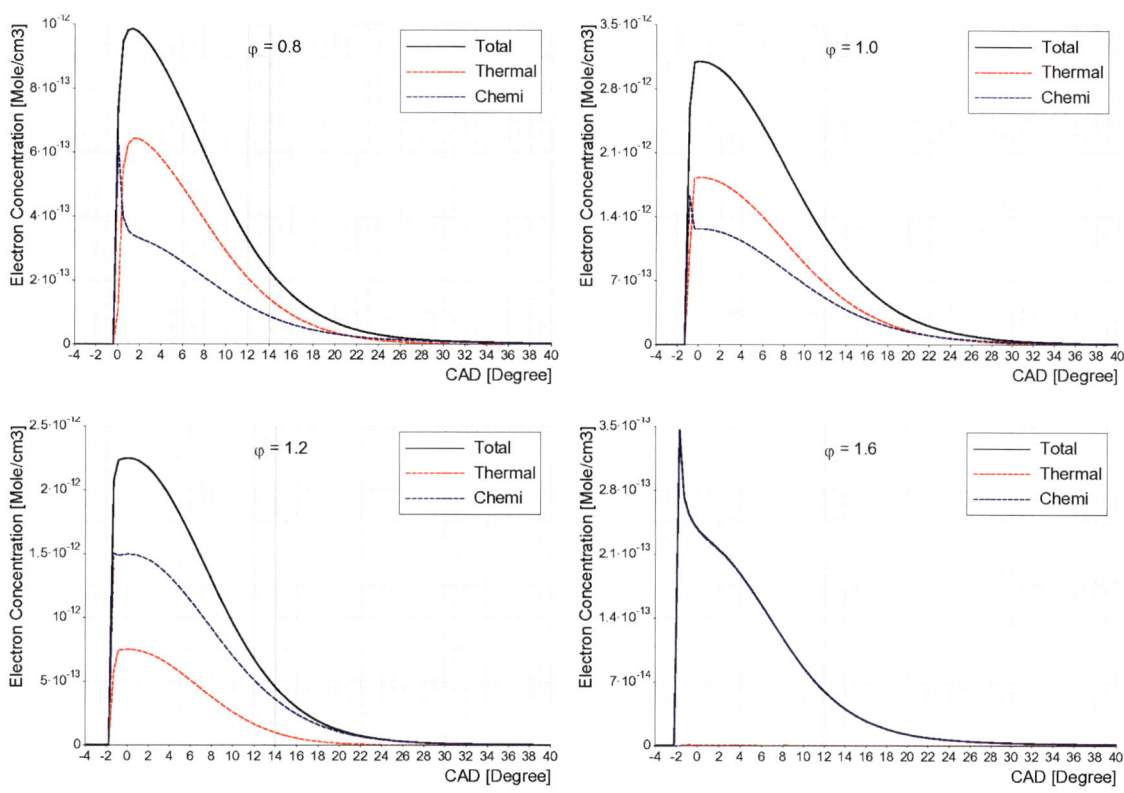

Figure 7.9   Breakdown of the total ion current signal at different equivalence ratios.

7.5 RELATIVE CONTRIBUTION OF CHEMICAL AND THERMAL IONIZATION OF NO

As previously discussed, NO^+ is a major contributor to the ion current in diesel engines. Therefore, the source of NO^+ formation is investigated. In SI engines, various studies showed that NO^+ is mainly formed by thermal ionization processes [5, 6, 86]. Chemical processes were ignored. NO mole fraction was assumed 3 to 10 times higher than the measured values in exhaust in order to match the calculated values with the experimental [60]. However, in diesel engines, the DIF model shows that NO^+ concentration in the combustion chamber is affected by thermal and chemical ionization processes. This agrees with the experimental data discussed in Chapter 6.

Figure 7.10 shows the effect of equivalence ratio on NO^+ formed by chemi-ionization (NO^+ Chemi) and by thermal ionization processes (NO^+ Thermal). Peak values of NO^+ Total, NO^+ Chemi, and NO^+ Thermal concentrations are shown in this figure.

Figure 7.10 indicates the following:

(a) NO^+ Total (Chemi + Thermal) exists between equivalence ratios 0.4 and 1.6.

(b) NO^+ Chemi is the only source of NO^+ formation between equivalence ratios 0.4 and 0.6.

(c)   The contribution of NO^+ Thermal starts at Φ higher than 0.6, and diminishes at Φ equals 1.6.

(d) NO^+ Thermal is dominant from equivalence ratio 0.8 to 1.4.

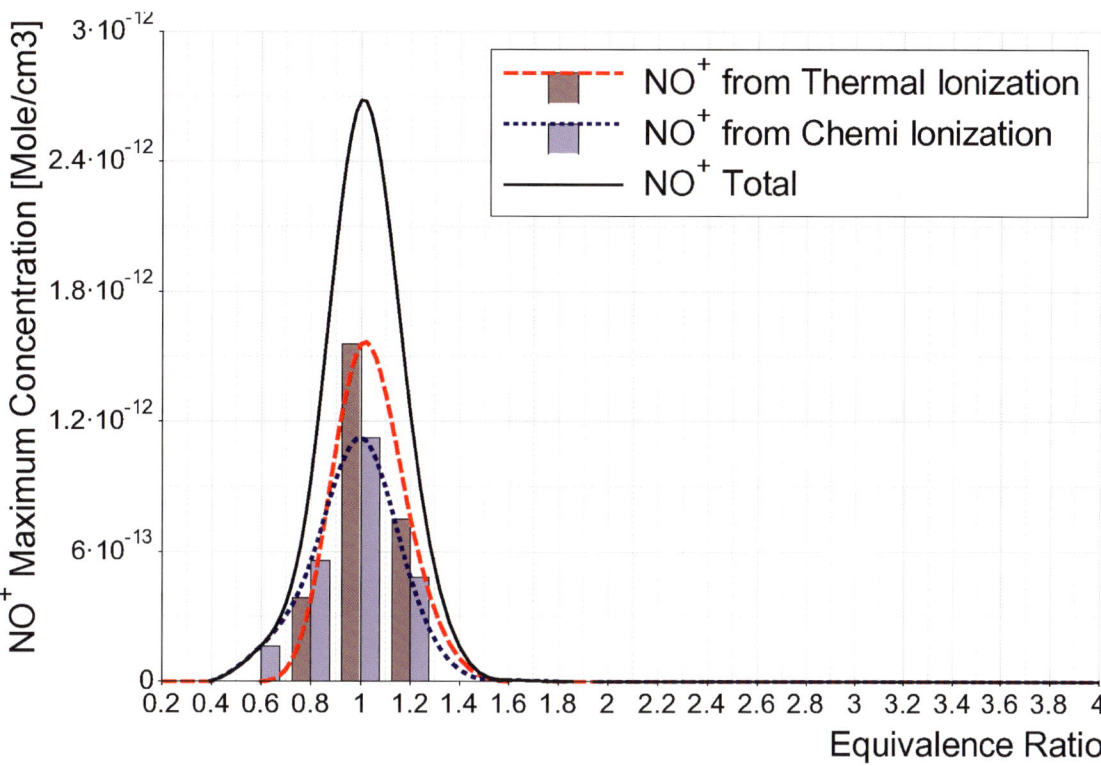

Figure 7.10   Peak values of NO^+ Total, NO^+ Chemi, and NO^+ Thermal concentrations at different equivalence ratios.

Figure 7.11 is a set of detailed graphs showing the breakdown of NO^+ at different equivalence ratios. (NO^+ Thermal) and (NO^+ Chemi) traces are plotted against crank angle degree. At Φ equal 0.4 and 0.6, only NO^+ Chemi exists. At higher equivalence ratios (Φ = 0.8, 1.0, 1.2, and 1.6), both NO^+ Chemi and NO^+ Thermal are present. A phase shift between the peak value of NO^+ Chemi and that of NO^+ Thermal is observed. This observation matches the results obtained experimentally and discussed previously in Chapter 6.

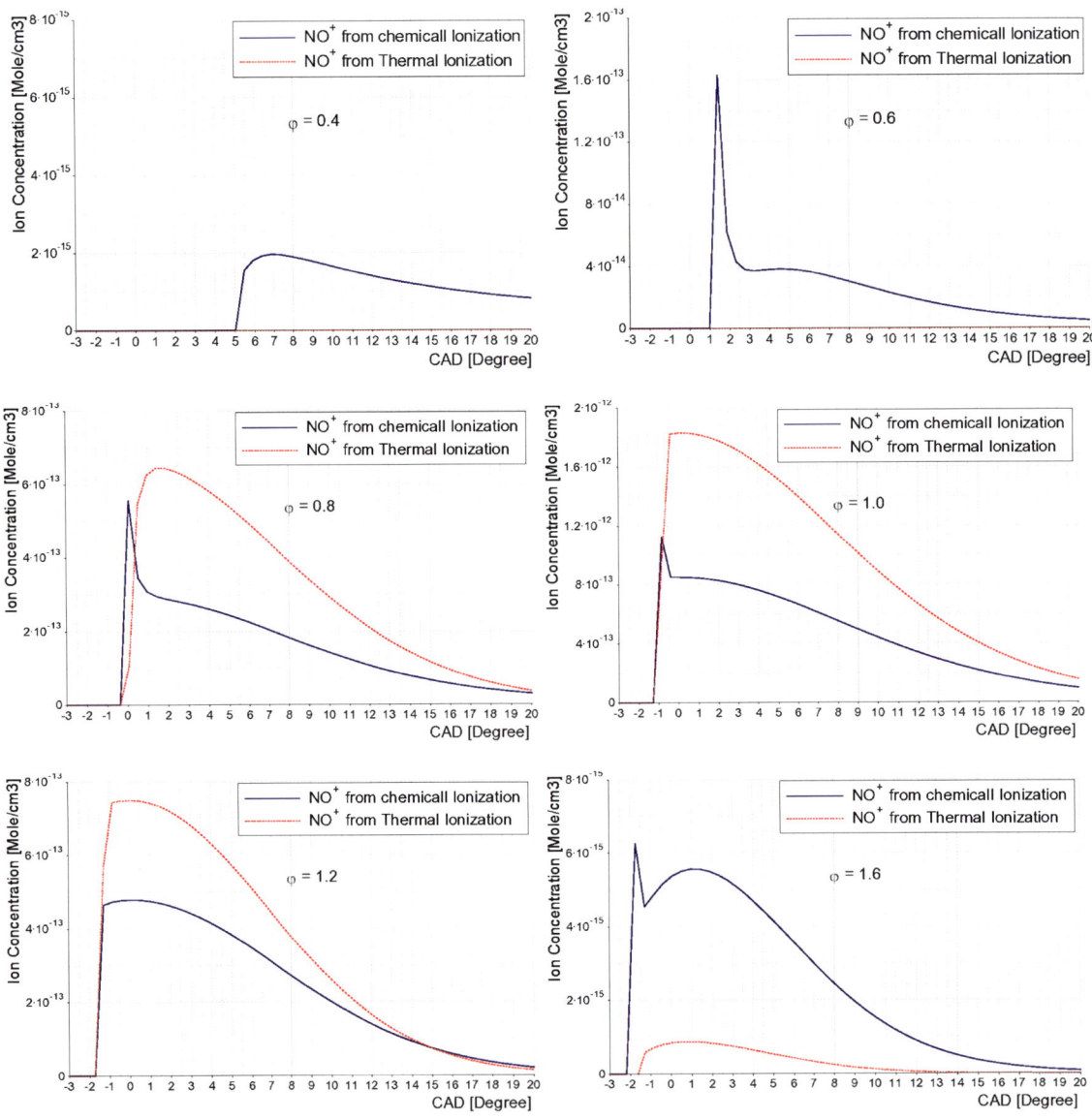

Figure 7.11   (NO$^+$ Thermal) and (NO$^+$ Chemi) traces plotted at different equivalence

ratios.

7.6 ION CURRENT SIGNAL BREAKDOWN

This section presents an analysis of an ion current signal calculated by the DIF model. Total ion current is divided into its two major components, thermal ionization and chemical ionization. In addition, ionized species contributing to each of these two components are shown.

Figure 7.12 shows the concentration of the different species which contribute in the ion current as a function of crank angle degrees during the combustion process at an equivalence ratio of 0.8. Total ion current signal is represented by the electron trace. Thermal ion current (Total Thermal) is composed of NO^+ Thermal as shown in the figure. Chemical ion current (Total Chemi) is composed of NO^+ Chemi and H_3O^+. It is noticed that the contribution of H_3O^+ in the total ionization is small compared to the contribution of NO. The concentrations of the remaining ions are insignificant at this equivalence ratio.   Total NO^+ is the addition of (NO^+ Thermal) and (NO^+ Chemi).

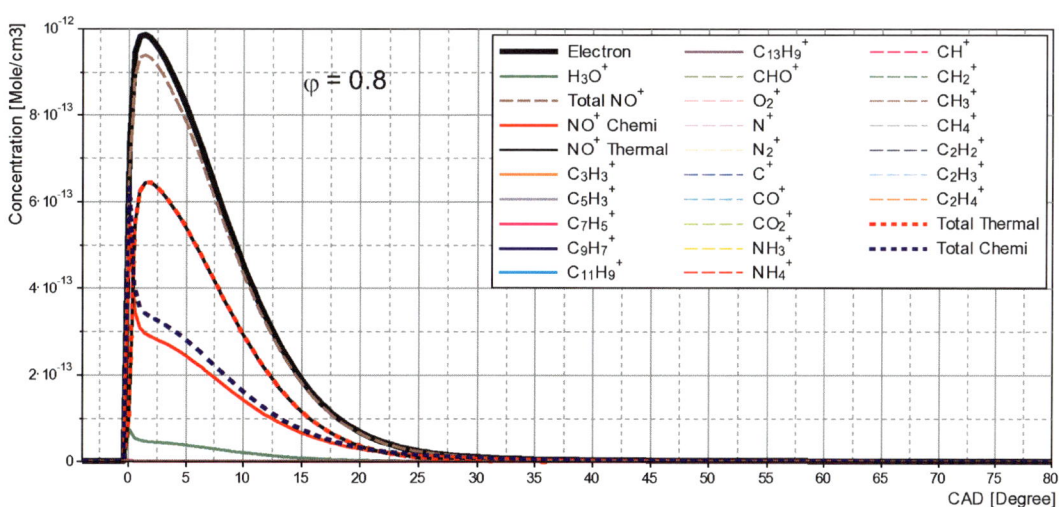

FIGURE 7.12   Sample of ion current signal breakdown at $\Phi = 0.8$.

7.7 COMARISON BETWEEN DIF AND EXISTING IONIZATION MODELS

Naoumov and Demin [16] developed a chemical kinetics model in 2004 to simulate the ion current in spark ignited engines. They used a total of 6 reactions (R6 to R11) including CHO+, H3O+, NO+, and electrons. Their model also included NO thermal ionization and chemi-ionization processes.

$$CH + O \leftrightarrow CHO^+ + e \qquad (R6)$$

$$CHO^+ + H_2O \leftrightarrow H_3O^+ + CO \qquad (R7)$$

$$CHO^+ + NO \leftrightarrow NO^+ + CHO \qquad (R8)$$

$$H_3O^+ + e \leftrightarrow 2H + OH \qquad (R9)$$

$$NO^+ + e \leftrightarrow N + O \qquad (R10)$$

$$NO + M \leftrightarrow NO^+ + e + M \qquad (R11)$$

Figure 7.13 gives a comparison between DIF model and Naoumov model simulation results at the same engine conditions. The simulation was conducted using Chemkin-Pro to show the differences between the two models at different equivalence ratios. Figure (7.13a) shows that both models predicted the same major ions (NO^+ and H_3O^+) contributing to the ion current signal at close to stoichiometry conditions. The difference in the amplitude of the two models is caused by NO^+ chemi. In Naoumov model, CHO^+ is consumed by reactions (R7 and R8) forming H_3O^+ and NO^+. However, in the DIF model, other reactions (R12 and R13) consuming CHO^+ are added. These reactions, with higher reaction rate, compete with reaction (R8) reducing the amount of NO^+ chemi formed.

$$CHO^+ + C_3H_2 \leftrightarrow C_3H_3^+ + CO \qquad (R12)$$

$$CHO^+ + CH_2 \leftrightarrow CH_3^+ + CO \qquad (R13)$$

Figures (7.13b, 7.13c, and 7.13d) show a clear discrepancy between the two models at richer mixtures which represent a higher load in a real engine. Naoumov model showed a dominance of H_3O^+ as the main source of ion current. The DIF model predicted different ionized species. DIF results are much closer to the experimental data discussed in Chapter 5 which shows that soot is a major contributor to the ion current signal at high engine loads.

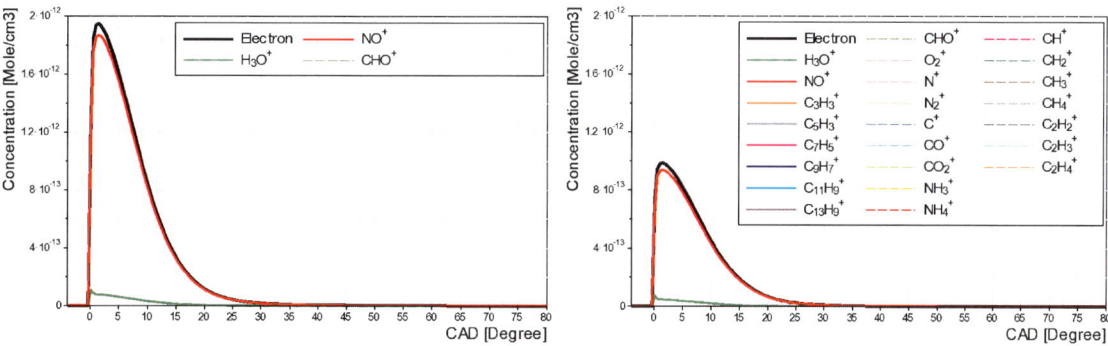

(a) Naoumov model (Left Graph) and DIF model (right Graph) results at Φ = 0.8

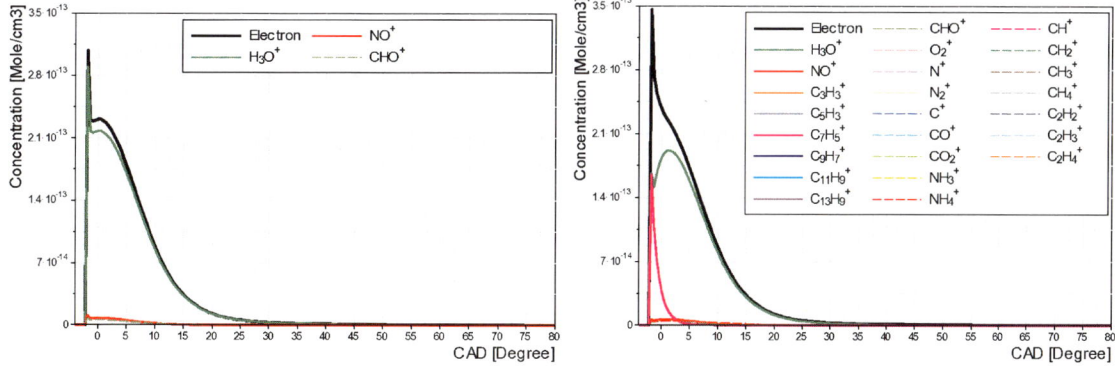

(b) Naoumov model (Left Graph) and DIF model (right Graph) results at Φ = 1.6

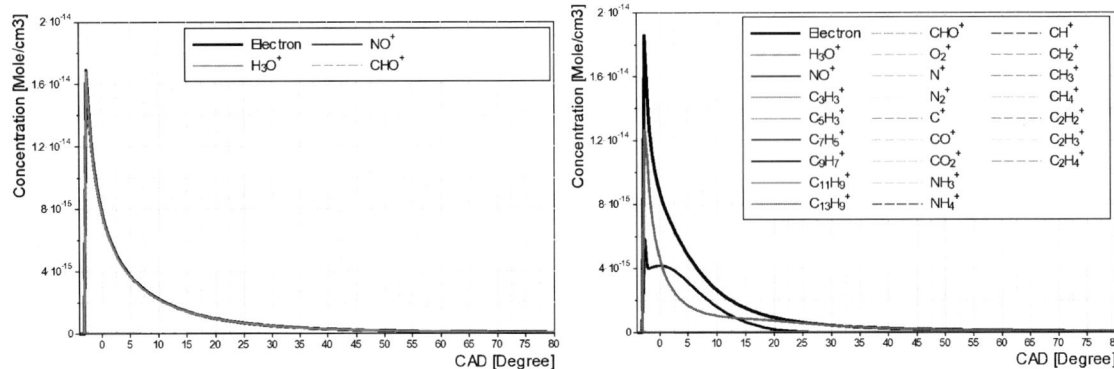

(c) Naoumov model and DIF model results at Φ = 3.0

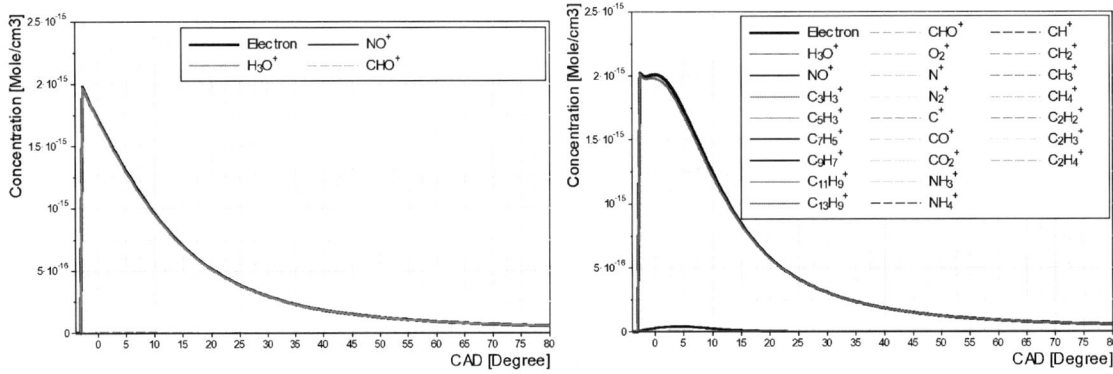

(d) Naoumov model and DIF model results at Φ = 4.0

Figure 7.13 Comparison between DIF model and Naoumov model at different equivalence ratios.

7.8 <u>CONCLUSION</u>

- A diesel ion formation (DIF) mechanism is developed consisting of a set of elementary reactions. These reactions include NO thermal and chemical ionization reactions.

- A zero-dimensional Chemkin model is used to study the effect of equivalence ratio on ions formed within a diesel cycle simulation.

- The model results showed that NO+ dominates the ion current signal at equivalence ratios around unity.

- A study was carried out to show the contribution of thermal and chemical ionization processes on the ion current. The study showed an equal contribution of both methods of ionization at certain equivalence ratios on the lean side.

- A complete ion current signal breakdown during the combustion process is presented showing the thermal and chemi ionization, including the contribution of hydrocarbon ions.

- A comparison between the results of the DIF model and a previous model presented by Naoumov was made.

CHAPTER 8   COMPUTATIONAL ANALYSIS

8.1 CHAPTER OVERVIEW

As discussed in the previous chapter, a chemical-kinetics mechanism (DIF) was developed using a zero-dimensional software (Chemkin) to simulate ion current in diesel engines. The mechanism considered the wide variability in the equivalence ratio for very lean, near stoichiometric, rich and soot producing mixtures.

In this chapter, the DIF model is introduced in a 3-D diesel cycle simulation computational fluid dynamics (CFD) code to determine the contribution of different species in the ionization process at different engine operating conditions. The CFD code is coupled with DARS-CFD, another module used to allow chemical kinetics calculations. The three-dimensional model accounts for the heterogeneity of the charge and the resulting variations in the combustion products. In addition, the model shows the effect of fuel injection pressure on the ion current characteristics. Ion current traces obtained from the John-Deere heavy duty diesel engine were compared to the 3-D model results.

8.2 DIESEL FUEL SURROGATE AND COMBUSTION MECHANISM

As commonly recognized, real fuels are complex mixtures of thousands of hydrocarbon compounds including linear and branched paraffins, naphthenes, olefins and aromatics. It is generally agreed that their behavior can be effectively reproduced by simpler fuel surrogates containing a limited number of components [116-118]. N-heptane as a diesel fuel surrogate is a large hydrocarbon molecule having a molecular weight very close to aviation fuels. Curran et al. (1998) developed a detailed n-heptane mechanism containing 570 species and 2520 reactions to study the oxidation of n-heptane in flow reactors, shock tubes and rapid compression machines [119]. In order to simulate diesel combustion with CFD codes, Curran's detailed mechanism was reduced by Reitz et al. (2004) at the Engine Research Center (ERC) in the University of Wisconsin-Madison [125]. The reduced n-heptane chemistry mechanism shown in APPENDIX A includes 33 species and 64 reactions including NO_x reaction mechanism. The developed diesel ion formation (DIF) model from Chapter 7 is used along with the University of Wisconsin reduced n-heptane mechanism to compute the concentrations of ions and intermediate combustion species in the diesel engine.

8.3 DIESEL CYCLE SIMULATION CODE

As a way to improve understanding of local combustion phenomena which could be hardly or impossibly detected from experimental techniques on a complete engine, CAE (Computer Aided Engineering) methods and in particular CFD (Computational Fluid Dynamics) gained an important role in the recent years. STAR-CD, a 3-D CFD code provided by CD-ADAPCO, is used to simulate the combustion process in direct injection (DI) diesel engines. However, a single CFD code is often unable to ensure a comprehensive analysis of complex physical systems. The coupling of different simulation tools, each specialized in different physical aspects, is therefore becoming more and more important both in industrial and university based research applications. DARS-CFD (Digital Analysis of Reaction Systems) is a tool used to simulate complex chemistry when coupled with CFD codes.

In addition, ES-ICE is an extra tool required for engine simulation. It automatically produces a parameterized meshed template that can be altered for specific engine configurations. ES-ICE produces a single mesh that can satisfy the full cycle of motion. This is possible because of the ability to add and remove layers of the mesh and change the connectivity of the mesh to maintain cell aspect ratios and quality throughout the entire cycle [126]. ES-ICE divides the simulated combustion chamber in equal sectors depending on the number of holes of the simulated fuel injector. This technique reduces the computational time significantly.

The model used in this research consists of a sector which represents the geometry of combustion chamber bowl of a diesel engine. The sector simulates fuel injection inside the combustion cylinder, and accounts for fuel atomization, evaporation, wall impingement, and swirl motion. In addition, the model considers the heterogeneity of the charge based on fuel and air mixing. Combustion is modeled after intake valve closing and before exhaust valve opening. The computational time is around 48 to 56 hours per case.

8.4 <u>CYCLE RESOLVED ANALYSIS OF IONIZATION IN DIESEL COMBUSTION</u>

The reduced N-Heptane combustion mechanism and the developed DIF model are applied to a diesel cycle simulation to investigate in details the contribution of the ionized species. A sector of the heavy duty John-Deere direct injection diesel engine (configurations listed in TABLE 3.1) with exact engine bowl geometry is used in this investigation. The sector model is shown in FIG 8.1. The effect of EGR (Exhaust Gas Recirculation) was not included in this study. The simulation covered a wide range of engine operating conditions including different loads and injection pressures.

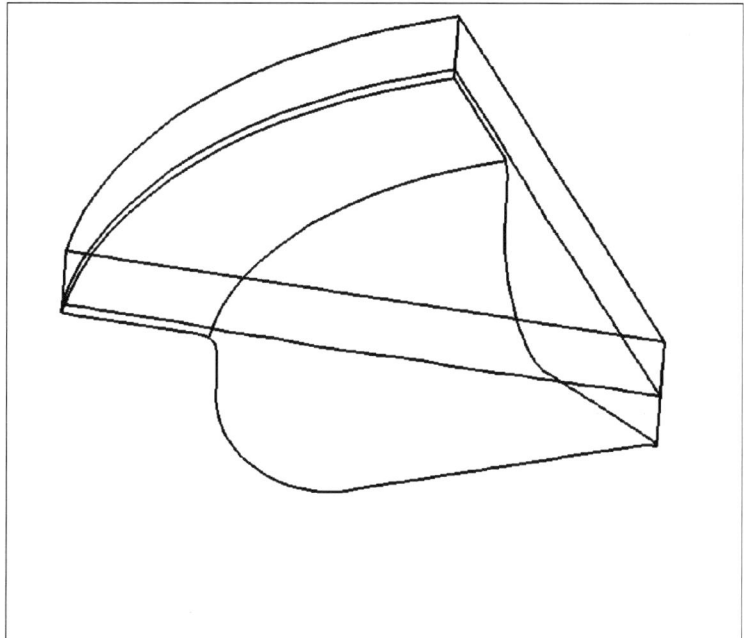

Figure 8.1   Sector of the combustion chamber bowl of the John-Deere direct injection diesel engine used in the 3-D model.

118

8.4.1 Experimental Results

Experiments were conducted to validate the 3-D model results. The test matrix shown in Table 8.1 consists of two injection pressures, 400 bar and 1000 bar. In each case, data such as cylinder pressure, ion current signal, in-cylinder NO concentration was recorded at low and high engine loads of 5 bar and 11 bar IMEP. The start of injection was kept constant at 8 CAD bTDC. A VGT was used to maintain the intake pressure constant at all engine loads and injection pressures. The in-cylinder gas sampling / ion current probe was fitted in the glow plug hole and used to detect the ion current signal during engine operation.

Figure 8.2 and FIG 8.3 show the pressure trace, needle lift signal, calculated temperature and rate of heat release at low and high loads for injection pressures of 400 bar and 1000 bar respectively.

TABLE 8.1

Injection Pressure	400 bar		1000 bar	
Engine Load	SOI	Fuel Flow (g/min)	SOI	Fuel Flow (g/min)
5 bar	8 bTDC	82.3	8 bTDC	75
11 bar	8 bTDC	197	8 bTDC	173

119

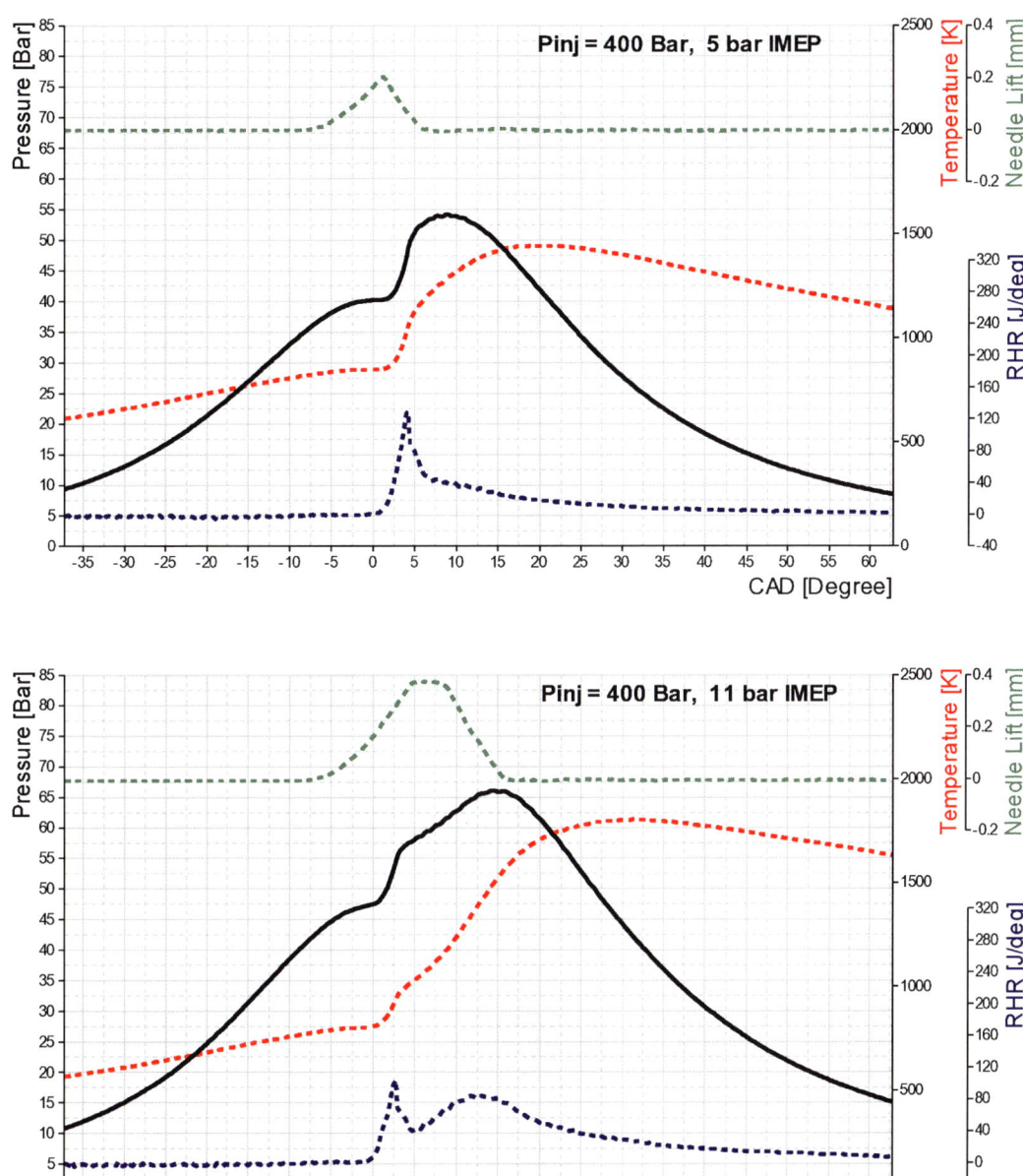

Figure 8.2   Cylinder gas pressure (Black line), needle lift (Green line), cylinder temperature (Red line), and calculated rate of heat release (Blue line) at 400 bar injection pressure for 5 bar IMEP (Top graph) and 11 bar IMEP (Bottom graph).

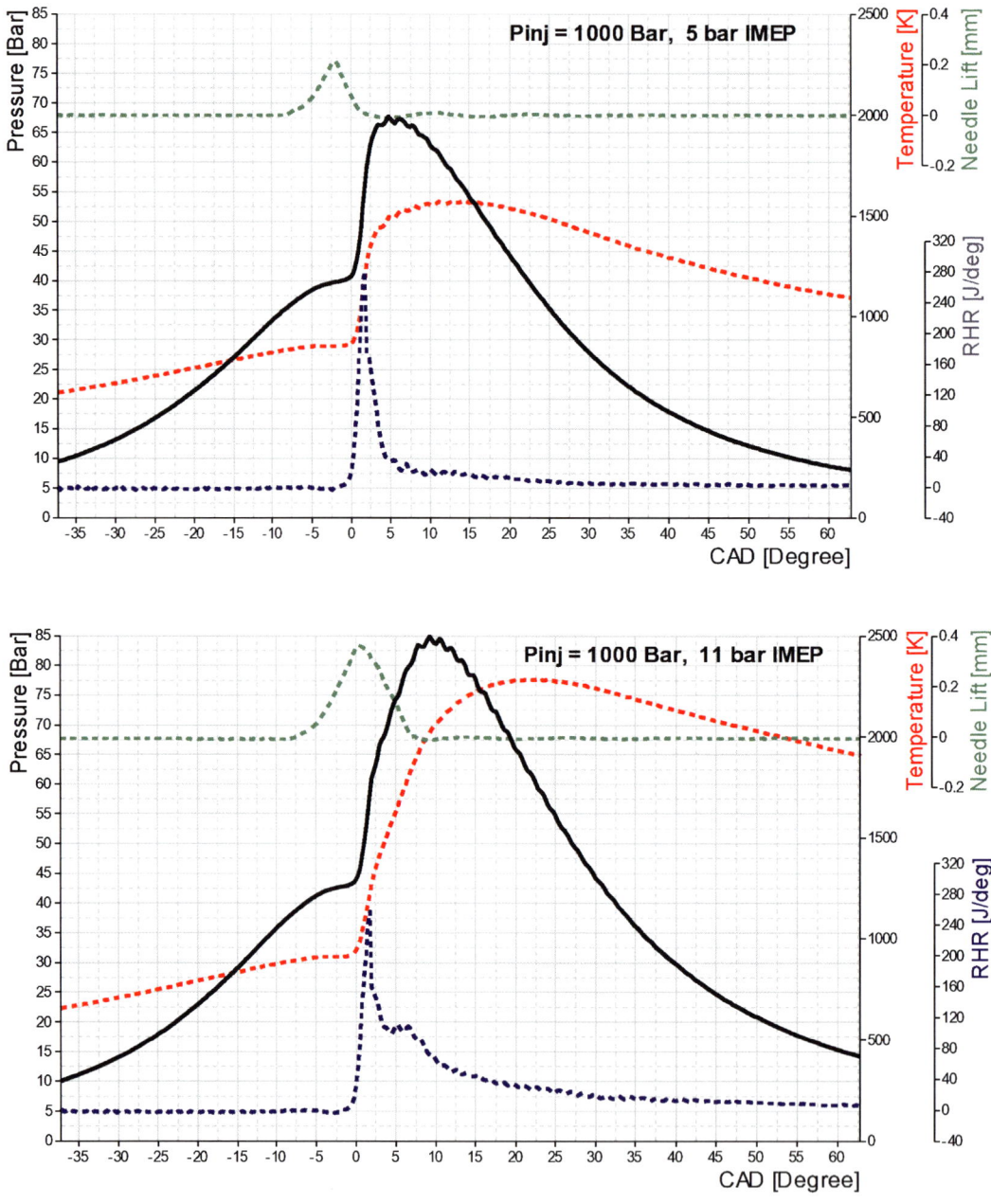

Figure 8.3   Cylinder gas pressure (Black line), needle lift (Green line), cylinder temperature (Red line), and calculated rate of heat release (Blue line) at 1000 bar injection pressure for 5 bar IMEP (Top graph) and 11 bar IMEP (Bottom graph).

8.4.2 Comparison between Experimental and Simulation Results

In this section, the results obtained from the experiments and the CFD STAR-CD 3-D model are compared at 4 engine operating conditions: 400 bar and 1000 bar injection pressures and a low load of 5 bar IMEP and a high load of 11 bar IMEP.

An assumption was made in the model in order to match the pressure traces obtained from the simulation with the experimental results. The start of injection (SOI) in the simulation has been retarded to 4 CAD bTDC instead of 8 CAD bTDC shown in Table 8.1. The reason is to account for the short ignition delay of n-Heptane.

Figure 8.4 is a representation of the rate of heat release (RHR) obtained in all 4 cases from experiments and simulation. As shown, fuel evaporation, premixed combustion, and diffusion controlled combustion are well predicted by the 3-D model using the reduced N-Heptane chemical kinetics model. There is a fairly good agreement between the experimental and modeling results.

Figures 8.5 and 8.6 show a comparison between the experimental data obtained in the heavy duty John-Deere engine of FIG 8.2 and FIG 8.3 and the predicted ion current traces from the 3-D model. Figure 8.5 represents a 100 cycle average of pressure traces and ion current signals obtained experimentally at 400 bar injection pressure with 5 bar and 11 bar IMEP. Simulation results of both traces are superimposed. Figure 8.6 shows the same traces as FIG 8.5 but for injection pressure 1000 bar. The figures show a fairly good agreement between the computed and measured ion current which reflects on the validity of the developed mechanism. More details about the composition of the ion current signal will be discussed in the coming

sections.

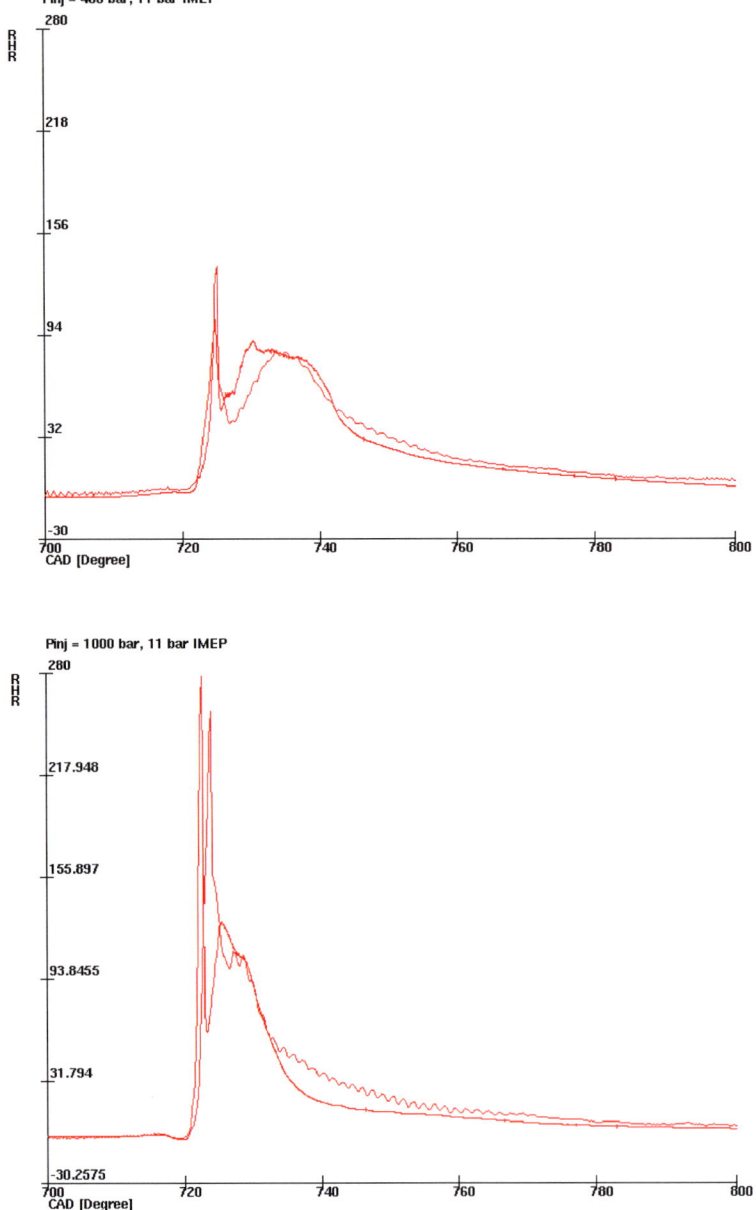

Figure 8.4   Comparison between experimental and predicted rate of heat release at

400 bar and 1000 bar injection pressure at 11 bar IMEP.

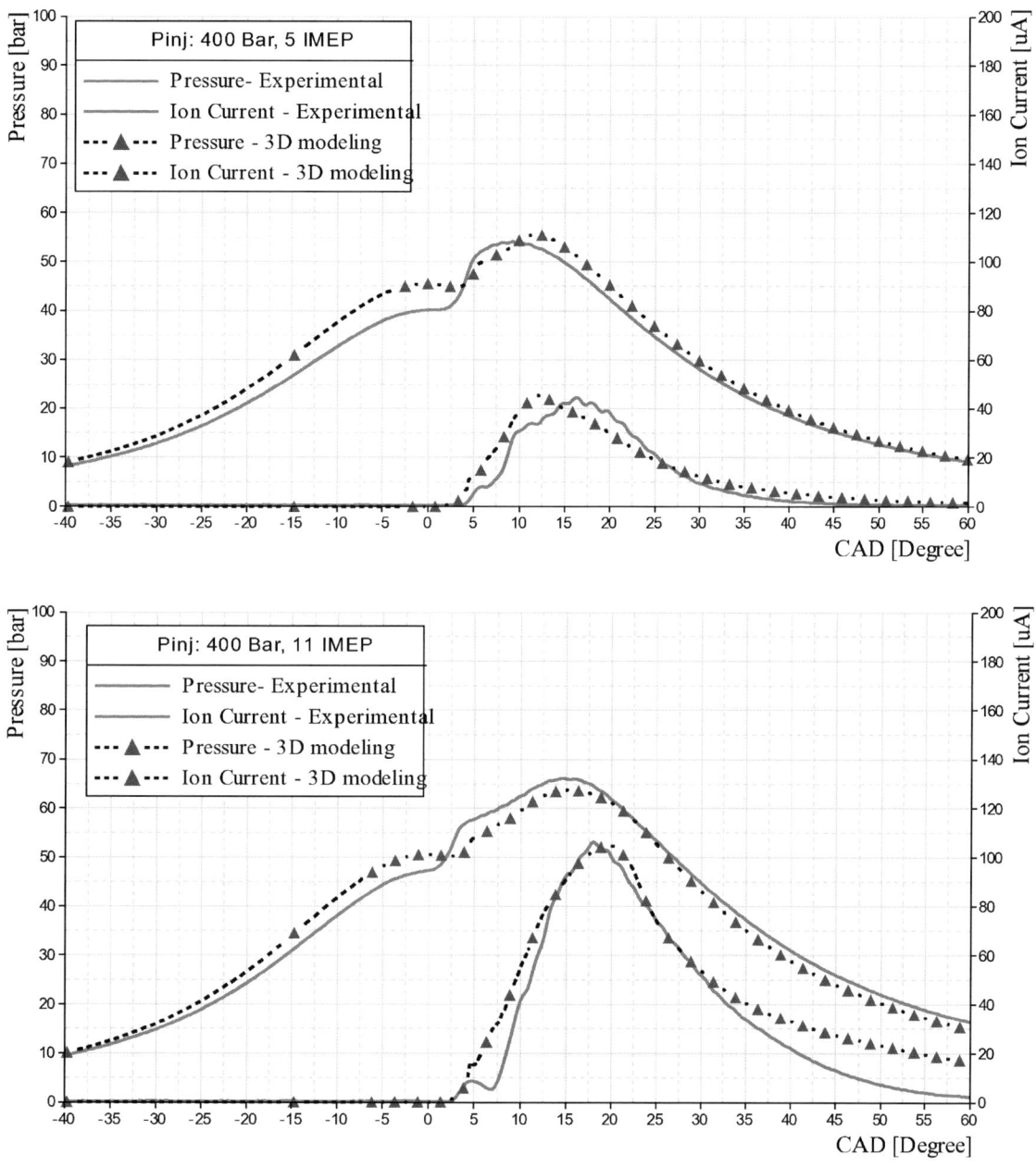

Figure 8.5   Comparison between experimental and predicted pressure traces and ion current for 400 bar injection pressure at 5 bar IMEP (Top graph) and 11 bar IMEP (Bottom graph).

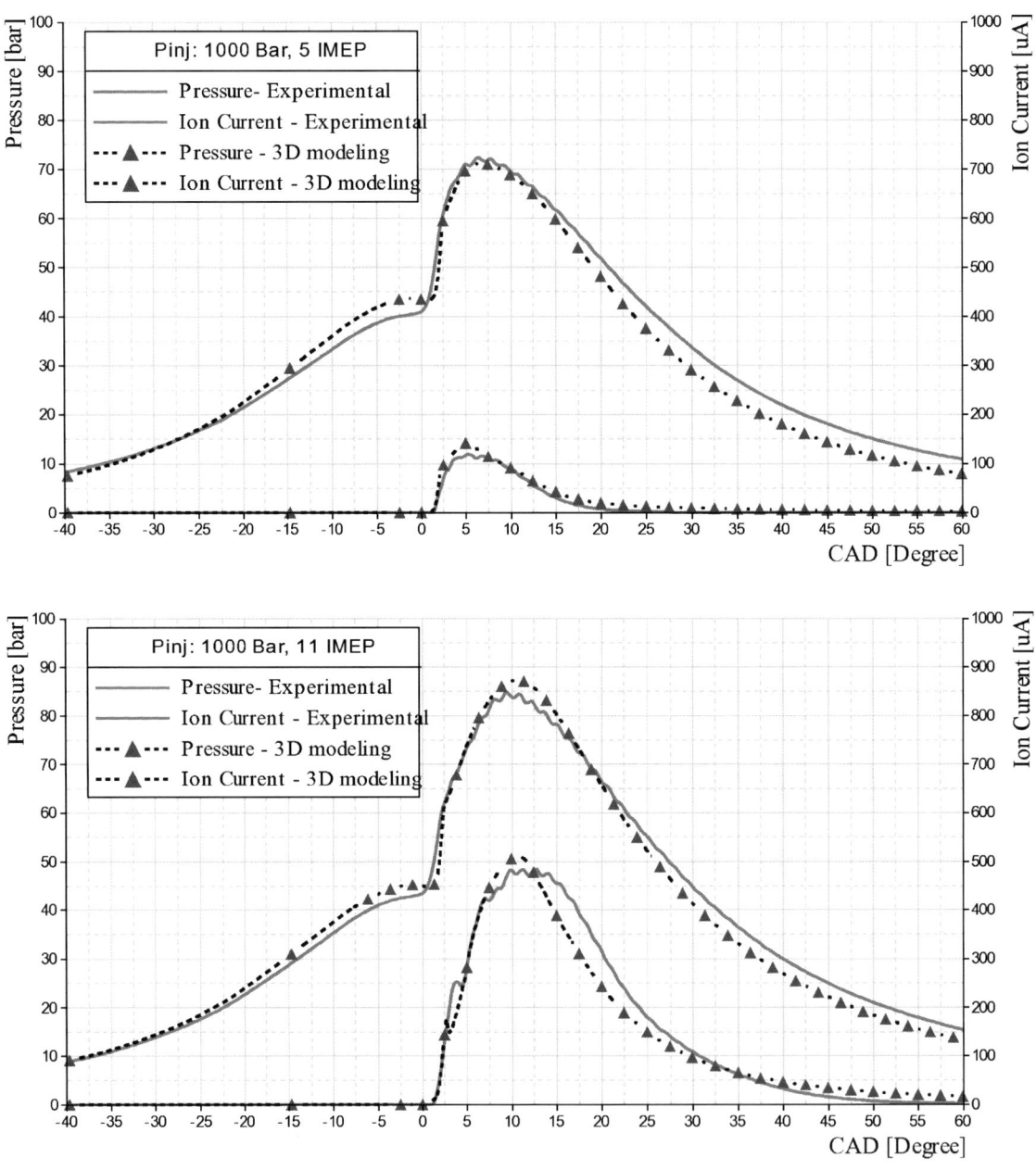

Figure 8.6   Comparison between experimental and predicted pressure traces and ion current for 1000 bar injection pressure at 5 bar IMEP (Top graph) and 11 bar IMEP (Bottom graph).

8.5 ION CURRENT SIGNAL BREAKDOWN

Analyzing and determining the major contributors into the ion current signal is essential for developing engine control strategies to meet stringent emissions standards and improved fuel economy. Figure 8.7 is a set of graphs obtained by the 3-D model showing the contribution of different species in the diesel ionization process at injection pressures 400 bar and 1000 bar. Based on the DIF model, only certain species showed dominance. The dominant ions have longer lifetime because they are rapidly formed by consuming other ions through charge transfer reactions. Moreover, they have low rate of consumption. The major contributors to the ion current signal in this diesel cycle simulation are H_3O^+, (NO^+ Total) represented by (NO^+ CHEMI) caused by chemical ionization processes and (NO^+ THERMAL) caused by thermal ionization. A small peak of $C_3H_3^+$ is shown in case of high injection pressure.

8.5.1 Effect of Injection Pressure

H_3O^+ obtained at 1000 bar injection pressure has higher amplitude than that obtained at 400 bar. In addition, H_3O^+ shows a clear bump following the first spike at low injection pressure. This will be discussed later in this chapter. As of NO^+, it reaches much higher values at high injection pressure. This is caused by a better fuel atomization thus more premixed combustion fraction and higher cylinder temperature and NO content.

8.5.2 Effect of engine load

Engine load has no effect on H_3O^+ first spike amplitude. As of NO^+, it does not show significant variation with engine load increase at low injection pressure. The reason behind this will be discussed in the following sections.

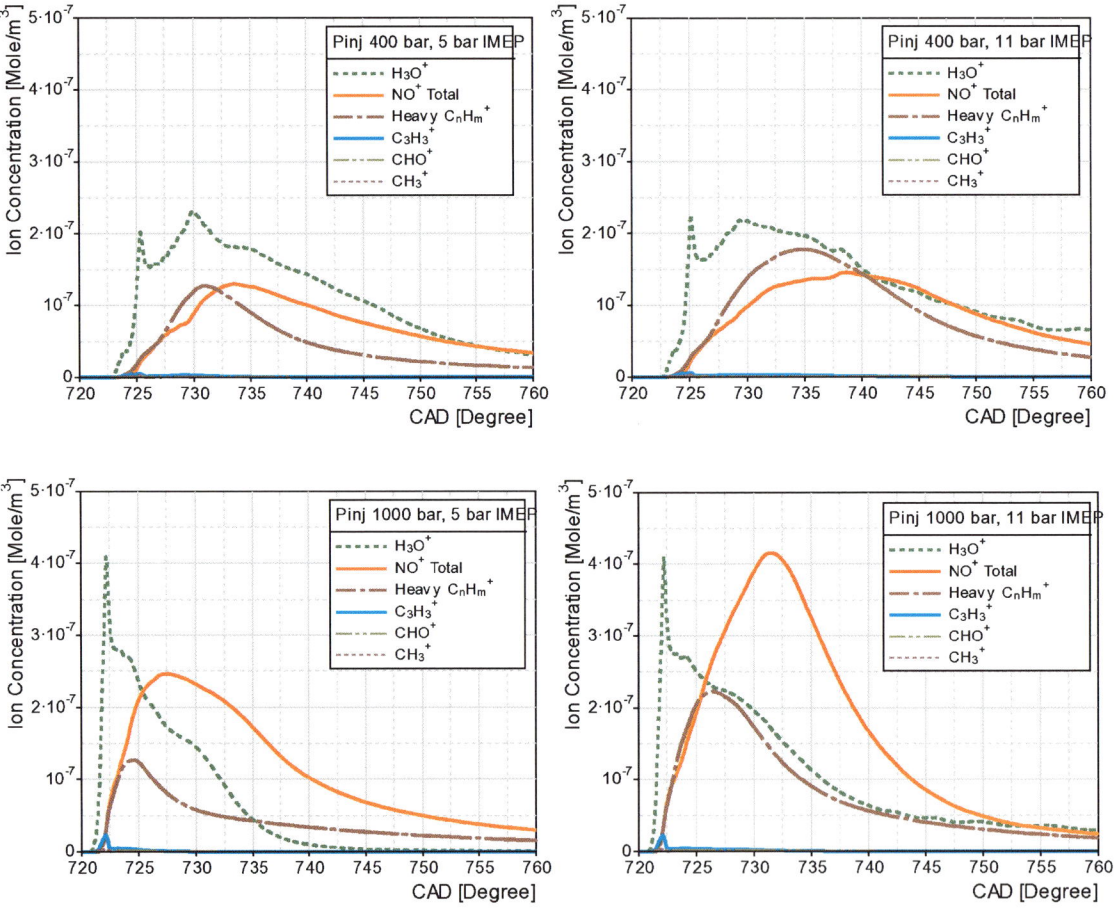

Figure 8.7 Detailed analysis of ion current carriers at 400 bar and 1000 bar injection pressure at low and high engine loads.

Figure 8.8 shows a detailed breakdown of the ion current signal of FIG 8.7 where the composition of NO^+ Total and $C_nH_m^+$ are represented at all 4 engine operating conditions. NO^+ CHEMI is significant at all engine loads and injection pressures while NO^+ THERMAL exists only at high injection pressure.

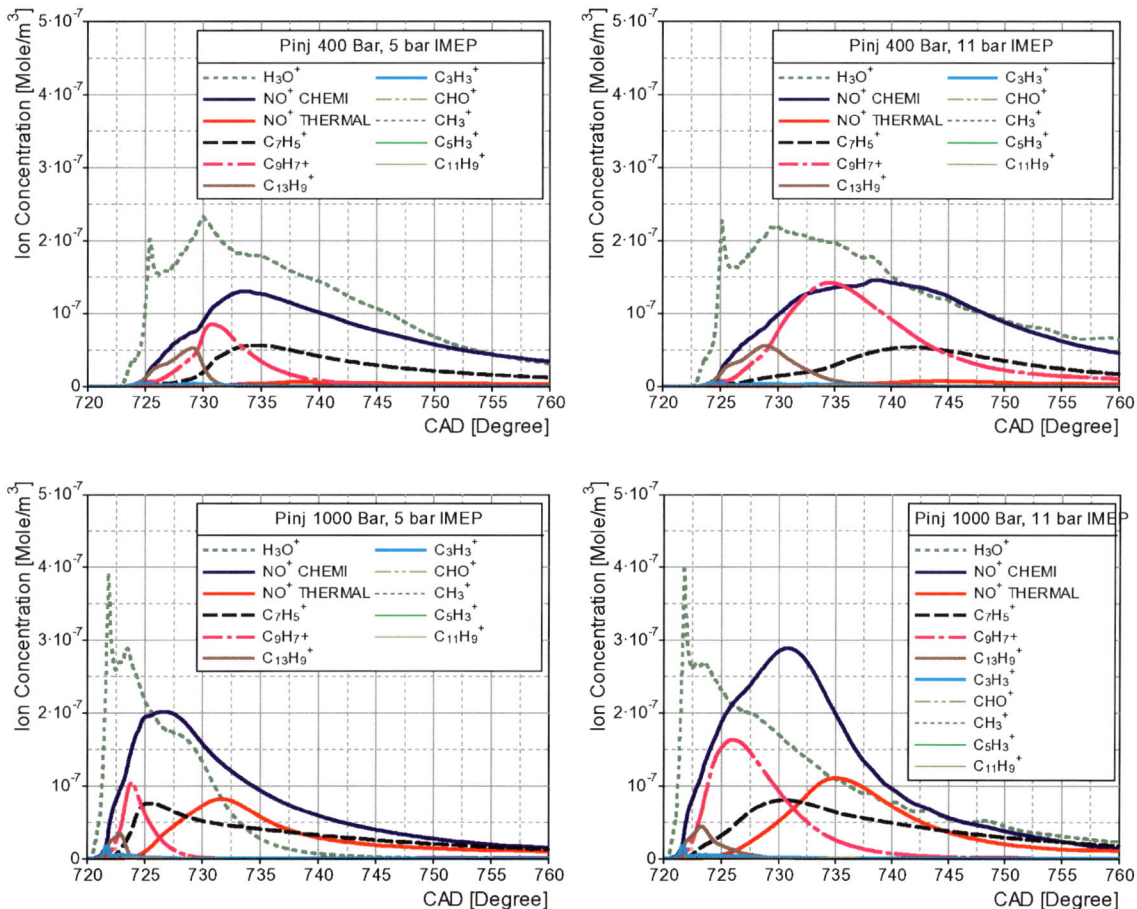

Figure 8.8   Detailed breakdown of the ion current signal based on the 3-D model results at different engine loads and injection pressures.

8.6 CORRELATION BETWEEN CH AND H_3O^+

The initiation reaction of ionization (R1) is based on forming CHO^+ out of CH and oxygen atoms. However, CHO^+ once formed is consumed by a charge transfer reaction (R2) forming H_3O^+ which is the first dominant and persistent ion. The correlation between CH and H_3O^+ is investigated.

$$CH + O \leftrightarrow CHO^+ + e \qquad (R1)$$

$$CHO^+ + H_2O \leftrightarrow H_3O^+ + CO \qquad (R2)$$

Figure 8.9 shows the 3-D model results for CH and H_3O^+ concentration at 400 bar and 1000 bar injection pressure, at low and high engine load. The figure shows a strong agreement between the two species. H_3O^+ highly depends on CH thus it follows the same trend. Tables 8.2 and 8.3 reflect a 3-D representation of CH and H_3O^+ distribution inside the engine bowl. The color code for all 3-D tables is shown in APPENDIX B.

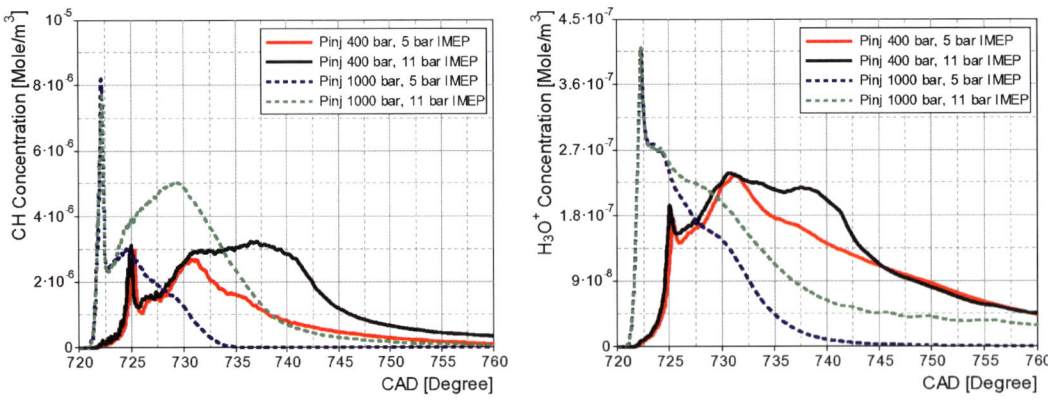

Figure 8.9   Traces obtained by 3-D model for CH concentration (Left graph) and H_3O^+ concentration (Right graph) at different engine operating conditions.

TABLE 8.2

CAD	Pinj = 400 bar, Load = 5 bar IMEP		Pinj = 400 bar, Load = 11 bar IMEP	
	CH	H_3O^+	CH	H_3O^+
723				
725				
730				
735				
740				

TABLE 8.3

CAD	Pinj = 1000 bar, Load = 5 bar IMEP		Pinj = 1000 bar, Load = 11 bar IMEP	
	CH	H_3O^+	CH	H_3O^+
723				
725				
730				
735				
740				

8.7 RELATIVE CONTRIBUTION OF CHEMICAL AND THERMAL IONIZATION OF NO

Figure 8.10 demonstrates a set of detailed graphs showing the breakdown of the total NO^+ content to its chemical and thermal components at low and high injection pressures and loads. As shown in the upper graphs, NO^+ THERMAL is negligible at low injection pressure at all loads, however, this is not the case at high injection pressure.

A comparison is made between two cases, (case 1: Pinj 400 bar at high load) and (case 2: Pinj 1000 bar and low load). The overall cylinder gas temperature of case 2 is slightly lower than that of case 1, however, NO^+ THERMAL in case 2 is much higher and reason behind this is the elevated local temperatures in case 2.

NO^+ THERMAL is observed to follow the temperature trace where both peaks occur relatively at the same location. In addition, a phase shift between NO^+ THERMAL and NO^+ CHEMI is noticed which is due to the fact that the main contributor to the thermal ionization of NO is the high temperature while chemical ionization of NO depends on charge transfer reactions between NO and other ionized species.

Tables 8.4, 8.5, 8.6, and 8.7 show a 3-D iso-surface representation of temperature, NO, NO^+ CHEMI, and NO^+ THERMAL inside a sector of the combustion chamber.

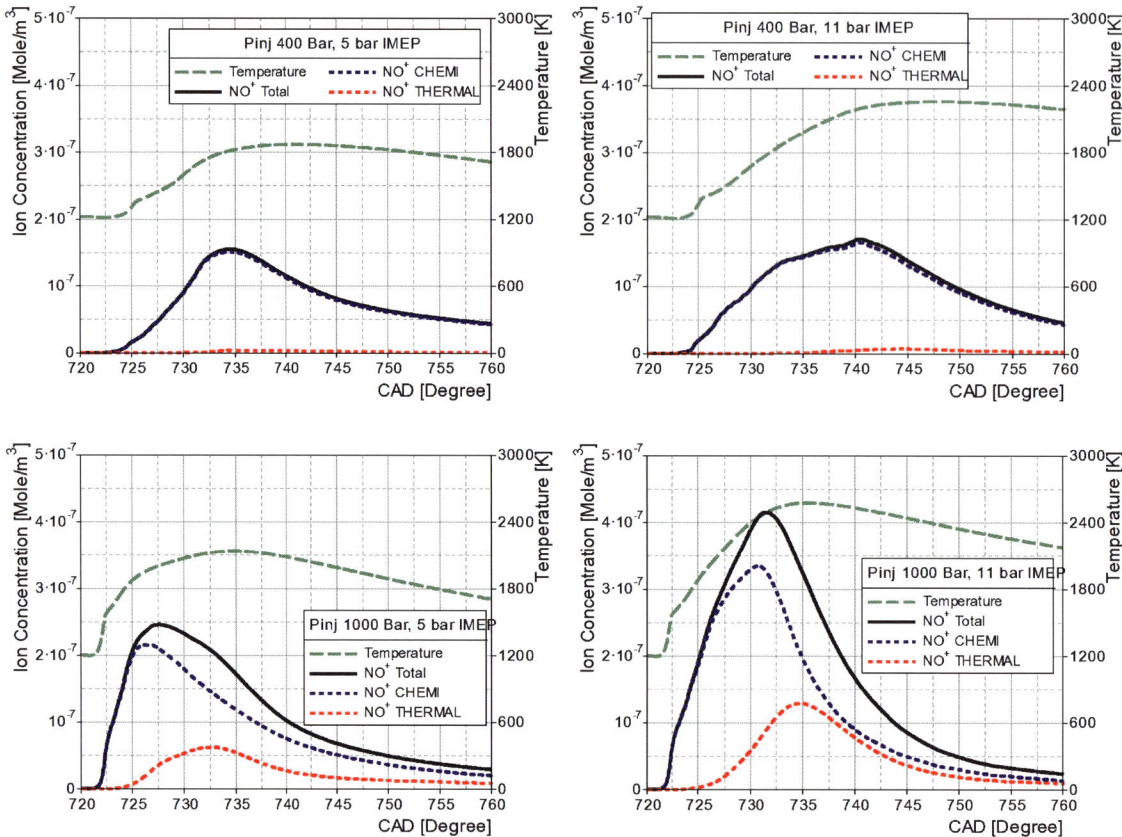

Figure 8.10   Temperature (Green line), NO$^+$ Total (Black line), NO$^+$ CHEMI (Blue line), and NO$^+$ THERMAL (Red line) traces obtained from 3-D modeling at low injection pressure (Top graphs) and high injection pressure (Bottom graph). 5 bar IMEP and 11 bar IMEP are simulated at each injection pressure.

TABLE 8.4

CAD	Pinj = 400 bar, Load = 5 bar IMEP			
	Temperature	NO	NO$^+$ Chemi	NO$^+$ Thermal
725				
730				
735				
740				
745				

TABLE 8.5

| CAD | Pinj = 400 bar, Load = 11 bar IMEP | | | |
	Temperature	NO	NO$^+$ Chemi	NO$^+$ Thermal
725				
730				
735				
740				
745				

TABLE 8.6

CAD	Temperature	NO	NO⁺ Chemi	NO⁺ Thermal
	Pinj = 1000 bar, Load = 5 bar IMEP			
725				
730				
735				
740				
745				

136

TABLE 8.7

| CAD | Pinj = 1000 bar, Load = 11 bar IMEP | | | |
	Temperature	NO	NO$^+$ Chemi	NO$^+$ Thermal
725	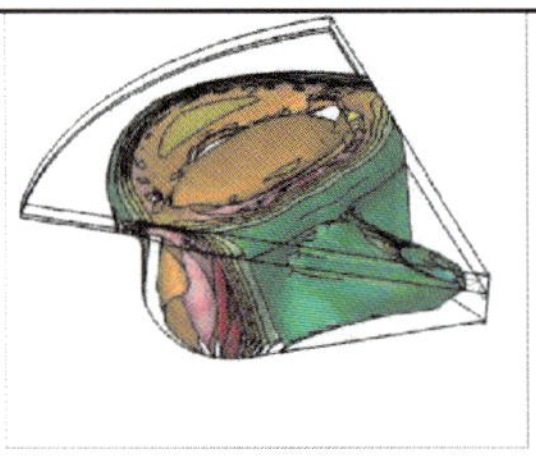	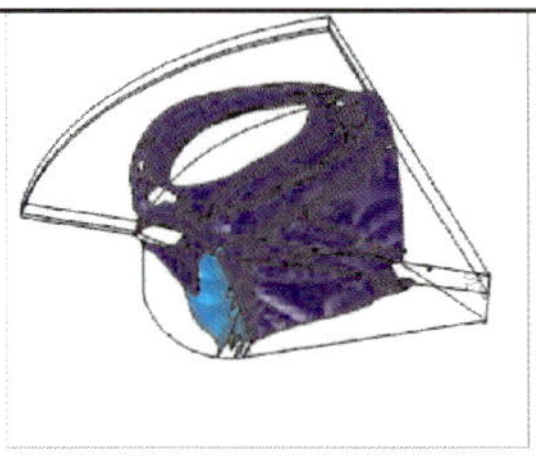	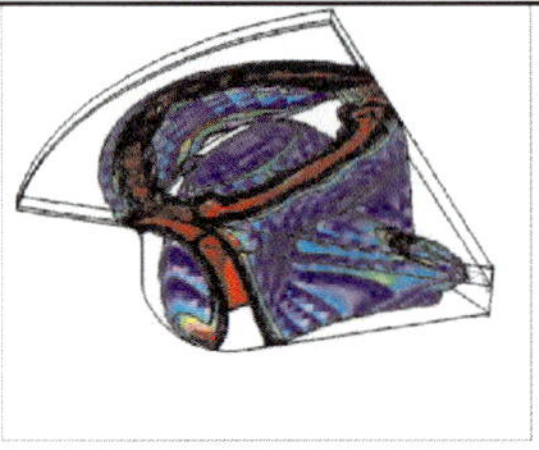	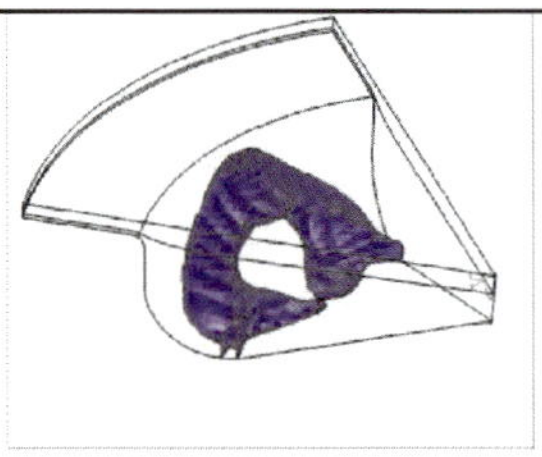
730	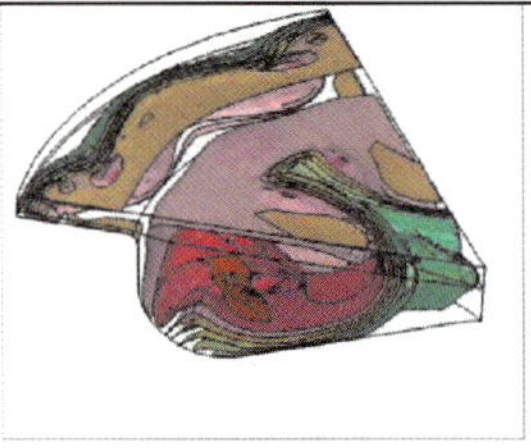	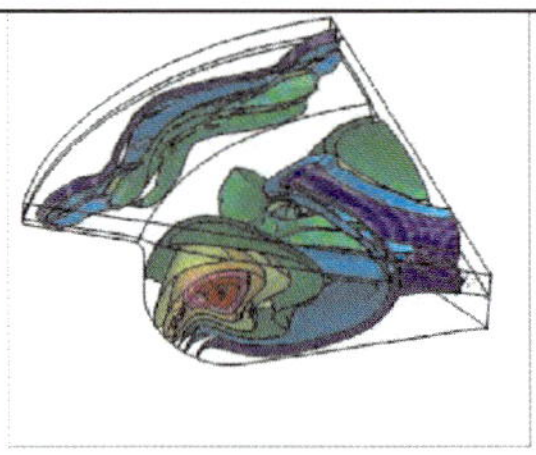	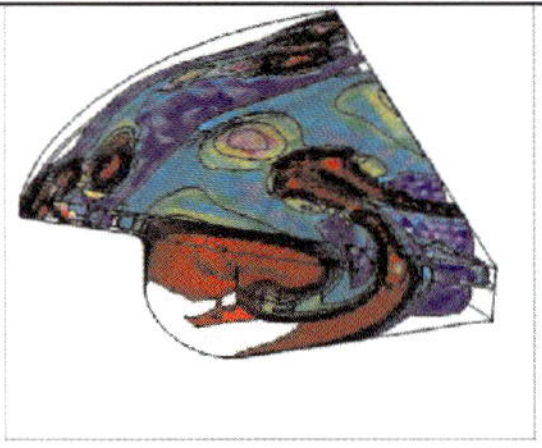	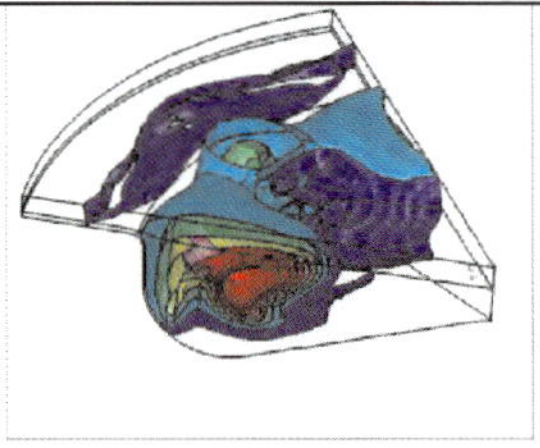
735	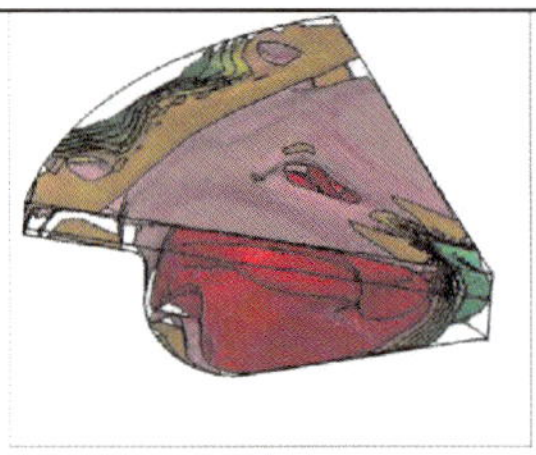	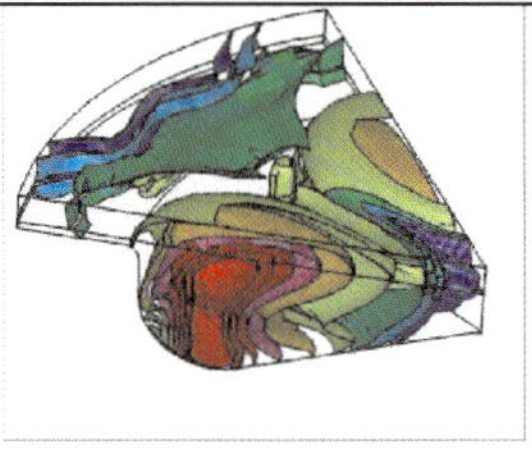	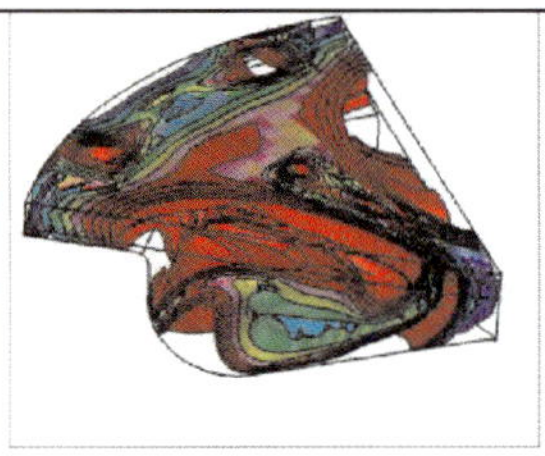	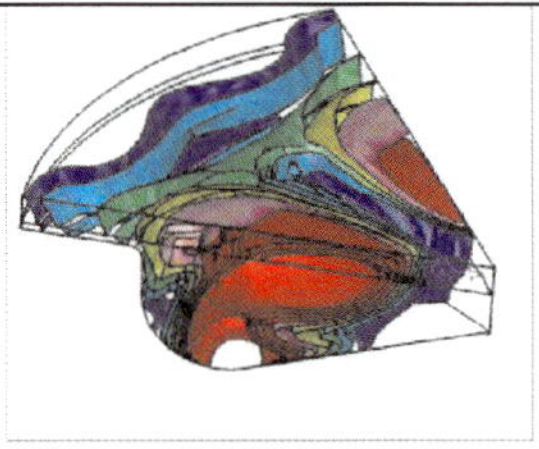
740	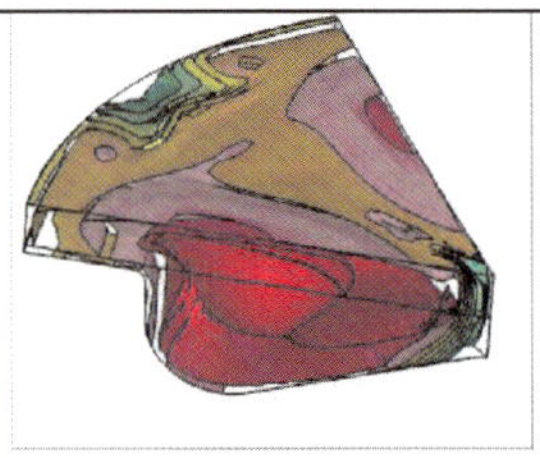	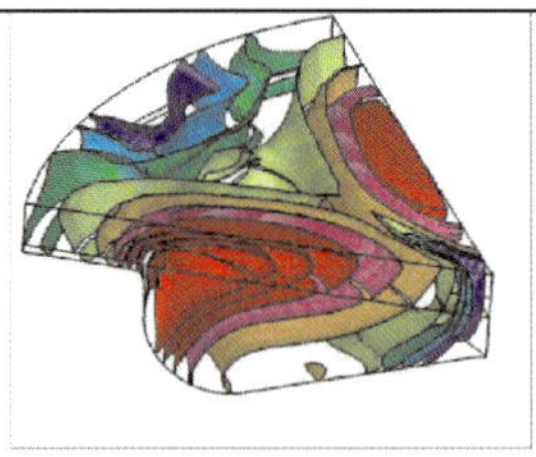	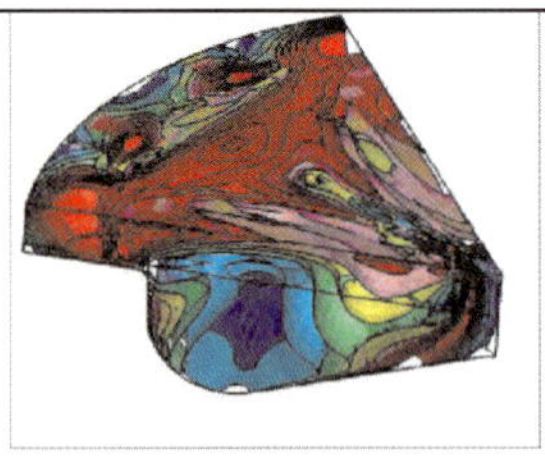	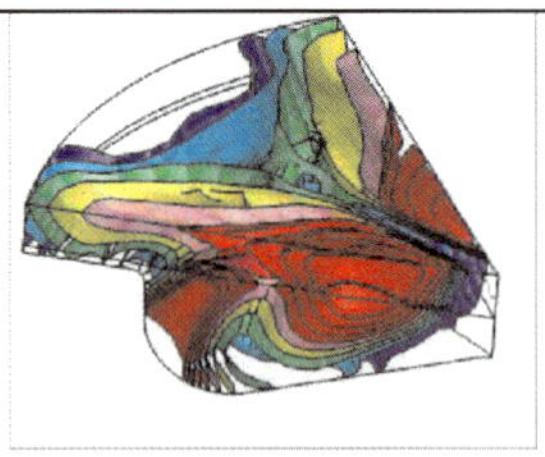
745	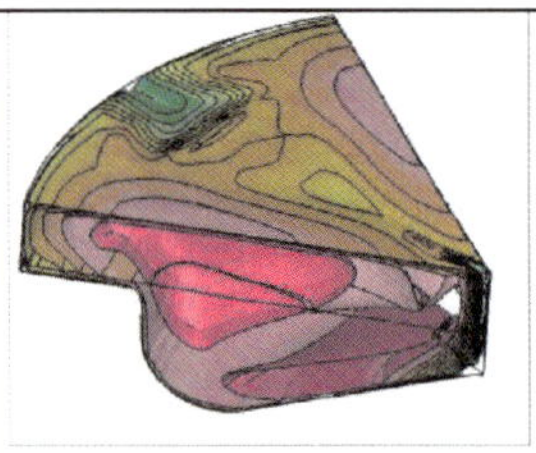	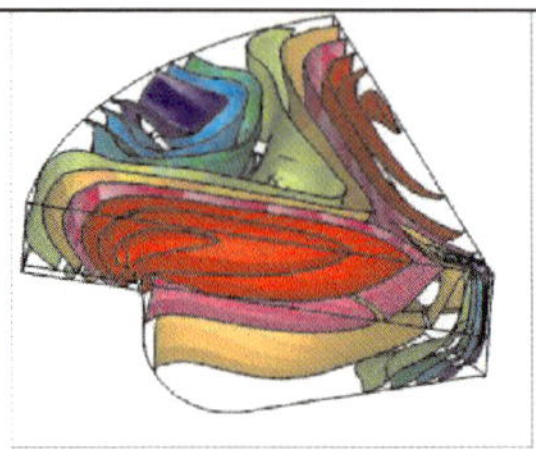	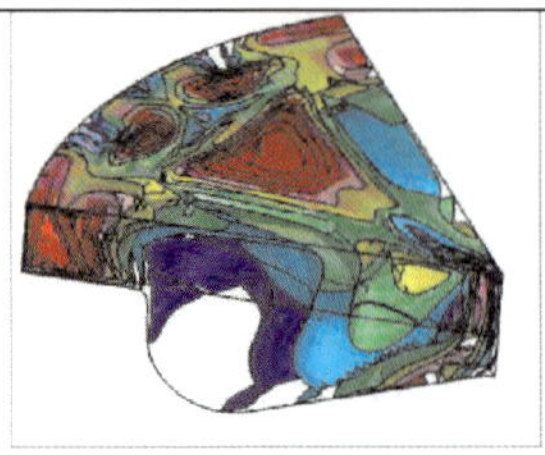	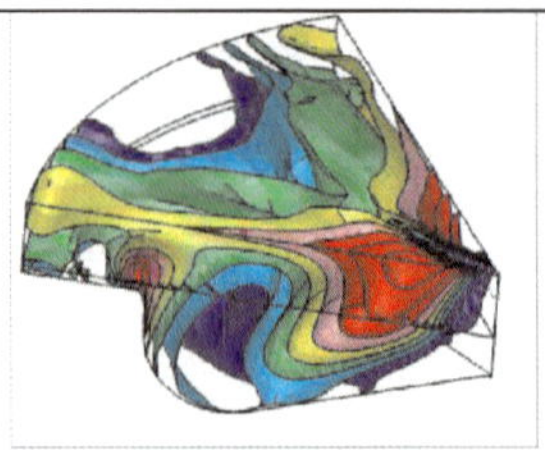

8.8 <u>CONCLUSIONS</u>

- The DIF model was applied in a 3-D CFD code to study the effect of injection pressure and engine load on the ion current signal. The model results agreed with the experimental data.

- Ion current due to NO thermal ionization is negligible at low injection pressures and low loads due to low local temperatures.

- 3-D representations are made for the distribution of different ions inside the combustion chamber at different engine loads and injection pressures.

CHAPTER 9 CONCLUSIONS

9.1 <u>CONCLUSIONS</u>

1. A new kinetic mechanism "Diesel Ion Formation, (DIF)" is developed for diesel engines. All published mechanisms in the literature are for ionization in spark ignition engines and HCCI, where the combustible mixture is homogeneous. DIF accounts for chemi-ionization and thermal ionization of NO and other species.

2. A new ion measuring technique is developed, referred to as "Multi Sensing Fuel Injector, (MSFI)". The fuel injector is used as an ion current probe, injection timing and a diagnostic tool for injection and combustion.

3. Another new technique is developed to use a gas sampling probe as an ion current sensor. Such a probe has been used in this dissertation to simultaneously measure the NO and ion current in a heavy duty diesel engines under different engine operating conditions.

4. A new technique is developed to predict the soot content in the exhaust from the ion current signal. This technique is applied to a heavy duty diesel engine and the predictions from the ion current signal agreed with soot measurements by an opacity meter, under steady and transient operating conditions.

5. A Chemkin zero-dimensional model is used to determine the effect of equivalence ratio on the ion current signal and the relative contribution of species such as H_3O^+, NO^+ in the ion current.

6. The DIF model is introduced to a 3-D CFD model to simulate the effect of injection pressure and engine load on the ion current signal. Experimental and modeling results showed a good agreement.

7. A 3-D representations of ionized species inside the combustion chamber showed the effect of injection pressure and load on the distribution of the ionized species and their variation during the combustion process.

APPENDIX A

Reactions in chemkin format		A	n	E
nc7h16 + h	= c7h15-2 + h2	4.380e+07	2.0	4760
nc7h16 + oh	= c7h15-2 + h2o	4.500e+09	1.3	1090
nc7h16 + ho2	= c7h15-2 + h2o2	1.650e+13	0.0	16950
nc7h16 + o2	= c7h15-2 + ho2	2.000e+15	0.0	47380
c7h15-2 + o2	= c7h15o2	1.560e+12	0.0	0.0
c7h15o2 + o2	= c7ket12 + oh	1.350E+14	0.0	18232.712
c7ket12	= c5h11co + ch2o + oh	3.530e+14	0.0	4.110e+4
c7h15-2	= c2h5 + c2h4 + c3h6	7.045E+14	0.0	3.46E+04
c3h7	= c2h4 + ch3	9.600e+13	0.0	30950.0
c3h7	= c3h6 + h	1.250e+14	0.0	36900.0
c3h6 + ch3	= c3h5 + ch4	9.000e+12	0.0	8480.0
c3h5 + o2	= c3h4 + ho2	6.000e+11	0.0	10000.0
c3h4 + oh	= c2h3 + ch2o	1.000e+12	0.0	0.0
c3h4 + oh	= c2h4 + hco	1.000e+12	0.0	0.0
ch3 + ho2	= ch3o + oh	5.000e+13	0.00	0.0
ch3 + oh	= ch2 + h2o	7.500e+06	2.00	5000.
ch2 + oh	= ch2o + h	2.500e+13	0.00	0.0
ch2 + o2	= hco + oh	4.300e+10	0.00	-500.
ch2 + o2	= co2 + h2	6.900e+11	0.00	500.
ch2 + o2	= co + h2o	2.000e+10	0.00	-1000.
ch2 + o2	= ch2o + o	5.000e+13	0.00	9000.

Reactions in chemkin format	A	n	E
ch2 + o2 = co2 + h + h	1.600e+12	0.00	1000.
ch2 + o2 = co + oh + h	8.600e+10	0.00	-500.
ch3o + co = ch3 + co2	1.570e+14	0.00	11800.
co + oh = co2 + h	8.987e+07	1.38	5232.877
o + oh = o2 + h	4.000e+14	-0.50	0.
h + ho2 = oh + oh	1.700e+14	0.0	875.
oh + oh = o + h2o	6.000e+08	1.30	0.
h + o2 + m = ho2 + m	3.600e+17	-0.72	0.
h2o2 + m = oh + oh + m	4.300e+16	0.00	45500.
h2 + oh = h2o + h	1.170e+09	1.30	3626.
ho2 + ho2 = h2o2 + o2	2.000e+12	0.00	0.
ch2o + oh = hco + h2o	5.563e+10	1.095	-76.517
ch2o + ho2 = hco + h2o2	3.000e+12	0.00	8000.
hco + o2 = ho2 + co	3.300e+13	-0.40	0.
hco + m = h + co + m	1.591E+18	0.95	56712.329
ch3 + ch3o = ch4 + ch2o	4.300e+14	0.00	0.
c2h4 + oh = ch2o + ch3	6.000e+13	0.0	960.
c2h4 + oh = c2h3 + h2o	8.020e+13	0.00	5955.
c2h3 + o2 = ch2o + hco	4.000e+12	0.00	-250.
c2h3 + hco = c2h4 + co	6.034e+13	0.0	0.0
c2h5 + o2 = c2h4 + ho2	2.000e+10	0.0	-2200.
ch4 + o2 = ch3 + ho2	7.900e+13	0.00	56000.

Reactions in chemkin format	A	n	E
oh + ho2 = h2o + o2	7.50E+12	0.0	0.0
ch3 + o2 = ch2o + oh	3.80E+11	0.0	9000.
ch4 + h = ch3 + h2	6.600e+08	1.60	10840.
ch4 + oh = ch3 + h2o	1.600e+06	2.10	2460.
ch4 + o = ch3 + oh	1.020e+09	1.50	8604.
ch4 + ho2 = ch3 + h2o2	9.000e+11	0.00	18700.
ch4 + ch2 = ch3 + ch3	4.000e+12	0.00	-570.
c3h6 = c2h3 + ch3	3.150e+15	0.0	85500.0
n+no<=>n2+o	3.500e+13	0.000	330.00
n+o2<=>no+o	2.650e+12	0.000	6400.00
n2o+o<=>no+no	2.900e+13	0.000	23150.00
n2o+oh<=>n2+ho2	2.000e+12	0.000	21060.00
n2o(+m)<=>n2+o(+m)	1.300e+11	0.000	59620.00
ho2+no<=>no2+oh	2.110e+12	0.000	-480.00
no+o+m<=>no2+m	1.060e+20	-1.410	0.0
no2+o<=>no+o2	3.900e+12	0.000	-240.00
no2+h<=>no+oh	1.320e+14	0.000	360.00
o+ch=h+co	4.000E+13	0.00	0.000E+00
Rev / 3.014E+14	0.22	1.752E+05 /	

Reactions in chemkin format	A	n	E
ch+o2=o+hco	3.000E+13	0.00	0.000E+00
Rev /	3.579E+13	0.01	7.183E+04 /
ch+co2=hco+co	3.400E+12	0.00	6.927E+02
Rev /	2.628E+08	0.90	6.362E+04 /
ch+h2o=oh+ch2	1.100E+12	0.00	-7.643E+02
Rev /	4.802E+10	0.24	-1.928E+04 /
ch+h2o=h+ch2o	4.600E+12	0.00	-7.643E+02
Rev /	4.828E+17	-0.91	6.019E+04 /
oh+ch=h+hco	3.000E+13	0.00	0.000E+00
Rev /	1.445E+16	-0.40	8.902E+04 /
h+ch2=ch+h2	6.000E+12	0.00	-1.791E+03
Rev /	2.147E+12	0.08	7.785E+02 /
ch3+M=ch+h2+M	6.900E+14	0.00	8.240E+04
Rev /	1.135E+10	0.91	-2.638E+04 /
ch+ch4=h+c2h4	3.000E+13	0.00	-4.060E+02
Rev /	5.974E+17	-0.86	5.918E+04 /
2CH2=C2H2+H2	1.200E+13	0.00	8.121E+02
Rev /	1.092E+20	-1.29	1.325E+05 /
2CH2=2H+C2H2	1.100E+14	0.00	8.121E+02
Rev /	3.779E+20	-1.34	2.830E+04 /

Reactions in chemkin format	A	n	E
O+C2H2=CH2+CO	2.000E+06	2.10	1.562E+03
	Rev / 5.924E-01	3.68	4.762E+04 /
C2H4+M=C2H2+H2+M	7.500E+17	0.00	7.949E+04
	Rev / 1.695E+12	1.14	3.469E+04 /
O+C3H4=CH2O+C2H2	1.000E+12	0.00	0.000E+00
	Rev / 2.244E+08	0.91	7.984E+04 /
C3H6=CH4+C2H2	3.500E+12	0.00	6.996E+04
	Rev / 1.417E+05	1.65	3.676E+04 /
CH+C2H2=H+C3H2	3.000E+13	0.00	-1.194E+02
	Rev / 1.722E+18	-1.02	3.792E+04 /
O+C3H2=C2H2+CO	6.800E+13	0.00	0.000E+00
	Rev / 8.925E+09	1.24	1.371E+05 /
OH+C3H2=HCO+C2H2	6.800E+13	0.00	0.000E+00
	Rev / 5.706E+11	0.62	5.098E+04 /

APPENDIX B

The color code for all 3-D tables (Table 8.1 to Table 8.13) in chapter 8.

Temperature	H3O+, NO+, C7H5+,	NO	CH
	C9H7+, C13H9+	Scale /1E5	Scale /1E10
	Scale /1E12		

2800.	800.0	800.0	800.0
2657.	746.4	746.4	746.4
2514.	692.9	692.9	692.9
2371.	639.3	639.3	639.3
2229.	585.7	585.7	585.7
2086.	532.1	532.1	532.1
1943.	478.6	478.6	478.6
1800.	425.0	425.0	425.0
1657.	371.4	371.4	371.4
1514.	317.9	317.9	317.9
1371.	264.3	264.3	264.3
1229.	210.7	210.7	210.7
1086.	157.1	157.1	157.1
942.9	103.6	103.6	103.6
800.0	50.00	50.00	50.00

REFERENCES

1. Henein, N. A., Bhattacharyya, A. , Schipper, J. , "Combustion and Emission Characteristics of a Small Bore HSDI Diesel Engine in the Conventional and LTC Combustion Regimes ," SAE Technical Paper 2005-24-045.

2. Henein, N. A., Badawy, T., Rai, N., and Bryzik, W., "Ion Current, Combustion and Emission Characteristics in an Automotive Common Rail Diesel Engine", ICEF2010-35123,Proceedings of the ASME 2010 Internal Combustion Engine Division Fall Technical ConferenceICEF2010.

3. http://www.epa.gov/airtrends

4. Peron, L., Charlet, A., Higelin, P., Moreau, B., and Burq, J. F., "Limitations of ionization current sensors and comparison with cylinder pressure sensors", SAE 2000-01-2830, SAE International, 2000.

5. Saitzkoff, A., and Reinmann, R., "An ionization equilibrium analysis of the spark plug as an ionization sensor," SAE Technical Paper 960337, 1996.

6. Saitzkoff, A., and Reinmann, R., "In cylinder pressure measurement using the spark plug as an ionization sensor," SAE Technical Paper 970857, 1997.

7. Auzins, J., Johansson, H., and Nytomt, J., "Ion-Gap Sense in Misfire Detection, Knock and Engine Control", SAE 950004, SAE International 1995.

8. Eriksson, L., Nielsen, L., and Nytomt, J., "Ignition control by ionization current interpretation", SAE 960045, SAE International, 1996.

9. Spicher, U., and Bäcker, H., "Correlation of flame propagation and in-cylinder pressure in a spark ignited engine" SAE 902126, SAE International, 1993.

10. Ohashi, Y., Fukui, W., and Ueda, A., "Application of vehicle equipped with ionic current detection system for the engine management system", SAE 970032, SAE International, 1997.

11. Eriksson, L., "Methods for Ionization Current Interpretation to be Used in Ignition Control", Licentiate thesis, Linköping University, Sweden, 1995.

12. Eriksson, L., Nielsen, L., and Glavenius, M., "Closed Loop Ignition Control by Ionization Current Interpretation," SAE 970854, 1997.

13. Kato, T., Akiyama, K., Nakashima, T., and Shimizo, R., "Development of Combustion Behavior Analysis Techniques in the Ultra Hign Engine Speed Range,: SAE 2007-01-0643, 2007.

14. Van Dyne, E., Burchmyer, C. L., Wahl, A. M. and Funaioli, A. E.," Misfire Detection from Ionization Feed Back Utilizing the Smartfire Plasma Ignition Technology," SAE Technical Paper 2000-01-1377, 2000.

15. Daniels, C. F., Zhu, G., G. and Winkelman, J.," Inaudible Knock and Partial-Burn Detection Using In-Cylinder Ionization Signal," SAE Technical Paper 2003-01-3149, 2003.

16. Wilstermann,A., Greiner,P., Hobner,R., Kemmler, R. R., and Schenk, J., "Ignition System Integrated AC Ion Current Sensing for Robust and Reliable Online Engine Control," SAE Technical Paper 2001-01-0555, 2000.

17. Abhijit, A. and Naber, J., "Ionization Response during Combustion Knock and Comparison to Cylinder Pressure for SI Engines," SAE Technical Paper 2008-01-0981, 2008.

18. Uparhyay, D., and Rizzoni, G., " AFR Control on a single-Cylinder Engine Using The Ionization Current \," SAE 980203,1999.

19. Ohashi, Y., Fukui, W., Tanabe, F. and Ueda, A.," The Application of Ion Current Detection System for the Combustion Limit Control," SAE Technical Paper 980171, 1998.

20. KLoevmark, H., Raek, P. and Forwell, U., "Estimating the Air/Fuel Ratio from Gausian Parameterization of the Ionization Currents in Internal Combustion SI Engines, SAE Technical Paper 2000-01-1215, 2000.

21. Rehim, A. A., Henein, N. A. and VanDyne, E, "Impact of A/F Ratio on Ion Current Features using Spark Plug with Negative Polarity," SAE Technical Paper 2008-01-1005, 2008.

22. Abhijit, A., Naber, J. and George, G.," Correlation Analysis of Ionization Signal and Air Fuel ratio for a Spark Ignition Engine," SAE 2009-01-0584, 2009.

23. Franke, A. R., Einwall, P., Johansson, B., Wickstrom, N., Reinmann, R. and Larsson, A., "The Effect of In-Cylinder Gas Flow on the Interpretation of the Ionization Sensor Signal," SAE Technical Paper 2003-01-1120, 2003.

24. Reinmann, R., and Saitzkoff, A., "Local air-fuel ratio measurements using the spark plug as an ionization sensor," SAE Technical Paper 970856, 1997.

25. Naoumov, V., and Demin, A., "Numerical study and analysis of pollutant production and emission control using ion current prediction in the SI engine," SAE Technical Paper 2003-01-0724, 2003.

26. Foster, J., Gunther, A., Ketterer, M., "Ion Current Sensing for Spark Ignition Engines", SAE Technical Paper 1999-01-0204, 1999.

27. Henein, N. , Bryzik, W. , Abdel-Rehim, A. , Gupta, A. , "Characteristics of Ion Current Signals in Compression Ignition and Spark Ignition Engines ," SAE Technical Paper 2010-01-0567.

28. Yoshiyama, S., Tomita, E., and Hamamoto, Y., "Fundamental study on combustion diagnostics using a spark plug as ion probe" SAE 2000-01- 2828, SAE International, 2000.

29. "Engine performance monitoring by means of the spark plug", Proceeding of the Institution of Mechanical Engineers, Volume 209, Part D2, P. 143-146, 1995.

30. "Engine performance monitoring using spark plug voltage analysis", Proceeding of the Institution of Mechanical Engineers, Volume 211, Part D6, P. 499-509

31. Eric, B., VanDyne, E., Wahl, A., Kenneth, R., and Lai, M., "In-cylinder air/fuel ratio approximation using spark gap ionization sensing", SAE 980166, SAE International, 1998.

32. Andersson I., and Eriksson L., "Ion sensing for combustion stability control of a spark ignited direct injected engine", SAE, 2000-01-0552, SAE International, 2000.

33. Franke, A., Einewall, P., Johansson, B., and Reinmann, R., "Employing an Ionization Sensor for Combustion Diagnostics in a Lean Burn Natural Gas Engine", SAE 2001-01-0992, SAE International, 2001

34. Beck, K., Gegg, T., Spicher, U., "Ion-Current Measurement in Small Two-Stroke SI Engines", SAE Technical paper 2008-32-0037, 2008.

35. Yoshiyama, S. , and Tomita, E. , "Combustion Diagnostics of a Spark Ignition Engine Using a Spark Plug as an Ion Probe ," SAE Technical Paper 2002-01-2838.

36. Kubach, H., Velji, A., Spicher, U., Fischer, W., "Ion Current Measurement in Diesel Engines", SAE Technical Paper 2004-01-2922, 2004.

37. Kurano, A. , "Glow Plug with Ion Sensing Electrode ," U.S. Patent 5 893 993, April 13, 1999.

38. Uhl, G. , "Ionic Current Measuring Glow Plug and Process and Circuit for its Activation ," U.S. Patent 6 549 013, April 15, 2003.

39. Girlando, S. , "Glow Plug Arranged for Measuring the Ionization Current of an Engine ," U.S. Patent 6 646 229, November 11, 2003.

40. Glavmo, M., Spadafora, P., Bosch, R., "Closed-loop start of combustion control utilizing ionization sensing in a diesel engine", SAE 1999-01-0549, SAE International, 1999.

41. Schweimer G., "Ion probe in the exhaust manifold of diesel engines", SAE 860012, SAE international, 1986.

42. Chae, J., Chung, S., and Jeong, Y., "A Study on Combustion Analysis of DI Diesel Engine Using Ionization Probe", JSAE 9531075, 1995.

43. Kessler, Michael, "Ionenstromsensorik im Dieselmotor," (Ph.D. thesis), Fortsch-Ber. VDI Reihe 12 Nr. 487, Dusseldorf.VDI Verog, 2002.

44. Jaegere, S., Deckers, J., and Van Tiggelen, "Identity of the most abundant ions in some flames," 8th Symposium (international) on Combustion, p. 155, 1962.

45. Bertand, C., and Van Tiggelen, "Ions in Ammonia Flames," J. Phys. Chem., Volume 78 No. 23, 1974.

46. Olson, D., and Calcote, H. F., "Ions in fuel-rich and sooting acetylene and benzene flames," Eighteenth Symposium (international) on Combustion, p. 453, 1981.

47. Calcote, H. F., and Keil, D. G., "The role of ions in soot formation," Pure &Appl. Chem., Vol. 62, No. 5, pp.815 – 824, 1990.

48. Hall-Roberts, V., Hayhurst, and Taylor, G., "The origin of soot in flames: Is the nucleus an ion?," Combustion and Flame, 120:578-584, 2000.

49. Hurle, I., and Sugden, M., "Chemi-ionization of nitric oxide in flames containing hydrocarbon additives," 12th Symposium (international) on Combustion, p. 387.

50. Debrou, Goodings, and Bohme, "Flame ion probe of intermediate leading to NO in CH4-O2-N2 flames," Combustion and Flame, Vol. 39 No. 1, Sept. 1980.

51. Lin, S., and Derek Teare, "Rate of ionization behind shock waves in air," The physics of fluids, Vol. 6 No. 3, March 1963.

52. Hansen, F. , "Ionization of NO at high temperature," Final Report on Phase I of NASA grant NAG-1-1211, Jan 1991.

53. Bulewicz, and Padley, "A study of ionization in cyanogen flames at reduced pressure by the cyclotron resonance method," 9th Symposium (international) on Combustion, p. 647.

54. Fialkov, and Kalinich, FGV, 1993.29, 111.

55. Goodings, M., and Bohme, D., "Detailed ion chemistry in Methane-Oxygen flames .II. Negative ions," Combustion and Flame, Vol. 36 No. 1, Sept. 1979.

56. Hayhurst, N., and Kittelson, D., "The positive and negative ions in oxy-acetylene flames," Combustion and Flame, Vol. 31 No. 1, 1978.

57. Brown, R., and Eraslan, A., "Simulation of ionic structure in lean close to stoichiometric acetylene flames," Combustion and Flame, Vol. 73 No. 1, July 1988.

58. Fialkov, A., "Investigations on ions in flames," Prog. Energy Combust. Sci. Volume 23, p. 399 – 528, 1997.

59. Eraslan, A. N., and Brown, R. C., "Chemiionization and ion-molecule reactions in fuel rich acetylene flames," Combustion and Flame, Vol. 74 No. 1, October 1988.

60. R Reinmann, R., and Saitzkoff, A., "Fuel and additive influence on the ion current," SAE Technical Paper 980161, 1998.

61. Naoumov, V., and Demin, A., "Modeling of combustion and non equilibrium ionization in spark ignition engines," SAE Technical Paper 2002-01-0009, 2002.

62. Mehresh, P., Souder, J., and Dibble, R. W. , "Combustion timing in HCCI engines determined by ion-sensor: experimental and kinetic modeling," Proceedings of the Combustion Institute Vol. 30, Issue 2, pp 2701-2709, Jan 2005.

63. Prager, J., Riedel, U., and Warnatz, J., "Modeling ion chemistry and charged species diffusion in lean methane-oxygen flames," Proceedings of the Combustion Institute Vol. 31, pp 1129-1137, 2007.

64. Badawy, T., "Investigation of the Ion Current Signal in Gen-Set Turbocharged Diesel", Master thesis Wayne State Univeristy, Detroit, USA, 2010.

65. Cambustion Ltd, "http://www.cambustion.com/products/cld500/cld-principles", Cambridge U.K., 2011.

66. Gupta, A., "Measurement and Analysis of Ionization Current Signal in a Single Cylinder Diesel Engine", Master Thesis Wayne State Univeristy, Detroit, USA, 2008.

67. Jansons, M. , Brar, A. , Estefanous, F. , Florea, R. , Taraza, D. , Henein, N. , Bryzik, W. , "Experimental Investigation of Single and Two-Stage Ignition in a Diesel Engine ," SAE Technical Paper 2008-01-1071.

68. Jansons, M. , Florea, R. , Zha, K. , Estefanous, F. ,Florea, E. , Taraza, D. , Henein, N. , Bryzik, W. , "Optical and Numerical Investigation of Pre-Injection Reactions and their Effect on the Starting of a Diesel Engine ," SAE Technical Paper 2009-01-0648.

69. Glavmo, M., Spadafora, P. and Bosch, R., "Closed Loop Start of Combustion Control Utilizing Ionization Sensing in a Diesel Engine," SAE Technical Paper 1999-01-0549, 1999.

70. Christensen,P., Bengtsson, J., Johansson, R., Vressner, A., Tunestal, P., Johansson, B., "Ion Current Sensing for HCCI Combustion Feedback", SAE Technical Paper 2003-01-3216, 2003.

71. Estefanous, F., "Multi-Sensing Fuel Injection System and Method for Making the Same", PCT Patent No. PCT/US2010/042549, July 2010.

72. Estefanous, F., Henein, N. A., "Multi Sensing Fuel Injector for Electronically Controlled Diesel Engines", SAE Technical Paper 2011-01-0936,2011.

73. Foster, J., Gunther, A., Ketterer, M., "Ion Current Sensing for Spark Ignition Engines", SAE Technical Paper 1999-01-0204, 1999.

74. Glavmo, M., Spadafora, P., Bosch, R., "Closed-loop start of combustion control utilizing ionization sensing in a diesel engine", SAE 1999-01-0549, SAE International, 1999.

75. Peron, L., Charlet, A., Higelin, P., Moreau, B., and Burq, J. F., "Limitations of ionization current sensors and comparison with cylinder pressure sensors", SAE 2000-01-2830, SAE International, 2000.

76. Badawy, T., Estefanous, F., Henein, N.A., "Simultaneous Ion Sensing and Gas Sampling in Combustion Systems", PCT Patent No. PCT/US2011/44141, 2011.

77. Cambustion Ltd, "In-Cylinder NOx Fast Sampling Technique", http://www.cambustion.com, 2011.

78. Estefanous, F., Badawy, T., Henein, N.A., "In-Cylinder Cycle-by-Cycle Soot Measuring Technique in Direct Injection Internal Combustion Engines," Provisional Patent Application No. 61447163, February 2011.

79. Henein, N. , Bryzik, W. , Abdel-Rehim, A. , Gupta, A. , "Characteristics of Ion Current Signals in Compression Ignition and Spark Ignition Engines ," SAE Technical Paper 2010-01-0567.

80. Badawy, T., Estefanous, F., Henein, N.A., "Simultaneous Ion Sensing and Gas Sampling in Combustion Systems", PCT Patent disclosure No. 11-1010, 2011.

81. Andersson, I., "A Comparison of Combustion Temperature Models for Ionization Current Modeling in an SI Engine," SAE Technical Paper 2004-01-1465, 2004.

82. Ahmedi, A., Franke, A., Sunden, B., "Prediction Tool for the Ion Current in SI Combustion", SAE Technical Paper 2003-01-3136, 2003.

83. Stenlaas, O., Johansson, B., "Measurement of Knock and Ion Current in a Spark Ignition Engine with and without NO Addition to the Intake Air", SAE Technical Paper 2003-01-0639, 2003.

84. Calcote, H. F., "Mechanisms for the formation of Ions in Flames", Combustion and Flame, Vol. 1, p. 385, 1957.

85. Amir Hossein, S., Ali, G., "Ion Current Simulation during the Post Flame Period in SI Engines", Iran J. Chem. and Chem. Eng., Vol. 24, No. 2, 2005.

86. Franke, A., Reinmann, R., Larsonn, A., "Analysis of the Ionization Equilibrium in the Post-Flame Zone", SAE Technical Paper 2003-01-0715, 2003.

87. Prager, J., Riedel, U., and Warnatz, J., "Modeling ion chemistry and charged species diffusion in lean methane-oxygen flames," Proceedings of the Combustion Institute Vol. 31, pp 1129-1137, 2007.

88. Albritton, D. L., "Ion-neurtral reaction-rate constants measured in flow reactors through 1977," Atomi data and nuclear data tables 22, 1-101, 1978.

89. Calcote, H. F., Gill, R., and Keil, D., "Modeling study to evaluate the ionic mechanism of soot formation in flames," Air Force Office of Scientific Research (approved for public release), AeroChem TP-518, 1993.

90. Lias, G., "Rate coefficients for ion-molecule reactions. Ions containing C and H," J. Phys. Chem. Ref. Data, Vol. 5, No. 4, 1976.

91. Meeks, E., Ho, P., "Modeling plasma chemistry for microelectronics manufacturing", Thin Solid Films 365 (2000) 334-347.

92. Gargo, S., Badawy, T., Estefanous, F., Henein, N. A., "Simulation of Ionization in Diesel Engines Premixed Combustion", ICARAME'11: International Conference on Advanced Research and Applications Mechanical Engineering Notre Dame University-Louaize, Lebanon, June 2011.

93. Vinckier, C., Gardner, M.P., Bayes, K.D., in 16thSymposium (International) on Combustion, The Combustion Institute, Pittsburgh, 1976, pp. 881-889.

94. Goodines, J.M., Bohme, D.K., Sueden, T.M., 16th Symposium (International) on Combustion, The combustion Institute, Pittsburgh, 1976, pp. 891-902.

95. Bascombe, K.N., Green, J.A., and Sugden, T.M., Adv. Mass Spectrometry, 1962, 2, 66.

96. Calcote, H.F., Combust. Flame, Sep 1981, Vol. 42, p. 215.

97. Calcote, H.F., and Keil, D.G., Combust. Flame, 1988, 74, 131.

98. Calcote, N.F., Olson, D.B., and Keil, D.G., Energy and Fuel, 1988, 2, 494.

99. Calcote, H.F., and Gill, R.J., "A Detailed Computer Model of Ion Growth in a Fuel Rich Acetylene/Oxygen Flame", Joint Meeting of the British and German Sections of the Combustion Institute, Queens College, Cambridge, England, 1993, p. 466.

100. Calcote, H.F., and Keil, D.G., Aero-Chem, TP-487, presented at III International Seminar on Flame Structure, Alma-A@ USSR, 18-22 September, 1989.

101. Calcote, H.F., and Keil, D.G., "Modeling Study to Evaluate the Ionic Mechanism of Soot Formation", AeroChem Research Lab, TP-569, 1998.

102. Calcote, H.F., Gill, R.J., and Keil, D.C., "Modeling Study to Evaluate the Ionic Mechanism of Soot Formation", AeroChem Research Lab, TP-518, 1993.

103. Calcote, H.F., Keil, D.C., Gill, R.J., and Berman, C.H., "Modeling Study to Evaluate the Ionic Mechanism of Soot Formation", AeroChem Research Lab, TP-531, 1994.

104. Francois, C., and van Tiggelen, P.J., Oxidation Combustion, 1980, 1, 163.

105. Bulewicz, E.M., and Padiey, P.J., in 9th Symposium (International) on Combustion, Academic Press, New York, 1963, pp. 647-658.

106. Bertrand, C. and van Tiggelen, P.J., J. Phys. Chem., 1974. 78, 2320.

107. Chase, M., "JANAF Thermochemical Tables Third Edition", Vol. 14, Supplement No. 1, 1985.

108. Chase, M., "NIST-JANAF Thermochemical Tables Fourth Edition", Journal of Physical and Chemical Reference Data, 1998.

109. McBride, B.J., Zehe, M.J., and Gordon, S., "NASA Glenn Coefficients for Calculating Thermodynamic Properties of Individual Species", NASA TP2002-211556, Sep 2002.

110. Draxl, K., Herron, J.T., "Energetics of Gaseous Ions", Journal of Physical and Chemical Reference Data, volume 6, 1977, Supplement No.1.

111. Journal of Physical and Chemical Reference Data, "Heat capacities and entropies of organic compounds in the condensed phase," Vol. 13, Supplement No. 1, 1984.

112. Sablier, M., and Fuji, T., "Mass Spectrometry of Free Radicals", Chem. Rev., 2002, 102 (9), 2855-2924.

113. Lias, S.G., Liebman, J.F., Levin, R.D., "Evaluated Gas Phase Basicities and Proton Affinities of Molecules; Heat of Formation of Protonated Molecules", J. Phys. Chem. Ref. Data, Vol. 13, No. 3, 1984.

114. Sarofim, A.F., Violi, A., "Quantum Mechanical Study of Molecular Weight Growth Process by Combination of Aromatic Molecules", Combustion and Flame 126:1506–1515 (2001).

115. Alberty, R.A., Reif, A., "Standard Chemical Thermodynamic Properties of Polycyclic Aromatic Hydrocarbons and their Isomer Groups I. Benzene Series", J. Phys. Chem. Ref. Data, Vol. 17, No. 1, 1988.

116. M. Mehl, H. J. Curran, W. J. Pitz and C. K. Westbrook, "Chemical kinetic modeling of component mixtures relevant to gasoline," European Combustion Meeting, Vienna, Austria, 2009.

117. Ra, Y., Reitz, R., "A reduced chemical kinetic model for IC engine combustion simulations with primary reference fuels", Combustion and Flame CNF:7021, 2008.

118. https://www-pls.llnl.gov/?url=science_and_technology-chemistry-combustion-n_heptane_version_3

119. Curran, H. J., P. Gaffuri, W. J. Pitz, and C. K. Westbrook, "A Comprehensive Modeling Study of n-Heptane Oxidation" Combustion and Flame 114:149-177 (1998).

120. R.J. Kee, F.M. Rupley, J.A. Miller, M.E. Coltrin, J.F. Grcar, E. Meeks, H.K. Moffat, A.E. Lutz, G. Dixon-Lewis, M.D. Smooke, J. Warnatz, G.H. Evans, R.S. Larson, R.E. Mitchell, L.R. Petzold, W.C. Reynolds, M. Caracotsios, W.E. Stewart, P. Glarborg, C. Wang, O. Adigun, W.G. Houf, C.P. Chou and S.F. Miller, Chemkin Collection, Release 3.7.1, Reaction Design, San Diego, CA (2003).

121. Seiser, H., H. Pitsch, K. Seshadri, W. J. Pitz, and H. J. Curran, "Extinction and Autoignition of n-Heptane in Counterflow Configuration," *Proceedings of the Combustion Institute* **28**, p. 2029-2037, 2000; Lawrence Livermore National Laboratory, Livermore, CA, UCRL-JC-137080.

122. T. J. Held, A. J. Marchese, and F. L. Dryer., "A semi-empirical reaction mechanism for n-heptane oxidation and pyrolysis", Combustion Science and Technology, 123:107–146,1997.

123. Tao, F., Reitz, R., Foster, D., " Revisit of Diesel Reference Fuel (n-Heptane) Mechanism Applied to Multidimensional Diesel Ignition and Combustion Simulations", Seventeenth International Multidimensional Engine Modeling User's Group Meeting at the SAE Congress, April 15, 2007, Detroit, Michigan.

124. Tanaka, S., Ayala, F., Keck, J.C., "A reduced chemical kinetic model for HCCI combustion of primary reference fuels in a rapid compression machine", Combustion and Flame 133 (2003) 467–481.

125. Patel, A., Reitz, R., "Development and Validation of a Reduced Reaction Mechanism for HCCI Engine Simulations", SAE Technical Paper 2004-01-0558, 2004.

126. CD-ADAPCO, http://www.cd-adapco.com/products/star_cd/index.html, 2011.

127. QuickField (Version 5.7), Computer Software, TeraAnalysis Ltd, Svendborg, Denmark.

128. NESS Engineering, "Technical data Field Enhancement Factor Equations", http://www.nessengr.com/techdata/fields/field.html, February 2009.

129. Calcote, H.F., Progress in Astronautics and Aeronautics, Vol. 12. Academic Press. Inc., New York, 1963, u. 107.

130. Calcote, H.F., in 8th Symposium (International) on Combustion, Williams and Wilkins, 1962, pp. 184-199.

131. Vinckier, C., Gardner, M.P., Bayes, K.D., in 16thSymposium (International) on Combustion, The Combustion Institute, Pittsburgh, 1976, pp. 881-889.

132. Peeters, J. and Vinckier, C., in 15th Symposium (International) on Combustion, The Combustion Institute, Pittsburgh, 1974, pp. 969-977.

133. Cool, T.A. and Tjossem, P.J.H., Chem. Phys. Len., 1984,111, 82.

134. Hou, Z. and Bayes, K.D., J. Phys. Chem., 1993, 97, 1896.

135. Goodines, J.M., Bohme, D.K., Sueden, T.M., 16th Symposium (International) on Combustion, The combustion Institute, Pittsburgh, 1976, pp. 891-902.

136.   Bascombe, K.N., Green, J.A., and Sugden, T.M., Adv. Mass Spectrometry, 1962, 2, 66.

137.   J.B. Heywood, Internal Combustion Engine Fundamentals, McGraw–Hill, New York (1988) p. 43.

138.   Michaud, P., Delfau, J.I., and Barassin, A., in 18thSymposium (International) on Combustion, The Combustion Institute, Pittsburgh, 198 1, pp. 443-45 1.

139.   Knewstubb, P.F., and Sugden, T.M., in 7th Symposium (International) on Combustion, Bunerworths, London, 1959, pp. 247-253.

140.   Goodings, J.M., Tanner, S.D., and Bohme, D.K., Can. J. Chem., 1982, 60, 2766.

141.   Vandooren, J., Mirapalheta, F. and van Tiggelen, P.J., "Study of the Influence of Nitrogen Oxides on the Chemi-ionization in C2H2/02 flames", The American Institute of Aeronautics and Astronautics, 1988, p. 104.

ABSTRACT

IONIZATION IN DIESEL COMBUSTION: MECHANISM, NEW INSTRUMENTATION AND ENGINE APPLICATIONS

by

FADI A. ESTEFANOUS

August 2011

Advisor: Dr. Naeim Henein

Major: Mechanical Engineering

Degree: Doctor of Philosophy

Diesel engines are known for their superior fuel economy and high power density. However they emit undesirable high levels of nitrogen oxide (NO_x) and black particulate smoke (Soot). To reduce these emissions, close loop engine control strategies are required. Therefore, there is a need for an in-cylinder combustion sensor. The ion current sensor has been used for combustion sensing in gasoline engines for which ionization mechanisms have been developed. This is not the case in diesel engines.

In this dissertation, a new mechanism for ionization in diesel engines has been developed and experimentally validated. Moreover, a three dimensional model has been implemented utilizing the new ionization mechanism to get more insight and better understanding of the ion current behavior in complex diesel combustion process. This model has been used to develop a new technique that allows the prediction of the soot content in the engine exhaust based on the ion current signal. Furthermore, a new arrangement has been applied for the use of the fuel injector as an ion current probe, injection timing sensor, and a diagnostic tool for injection and combustion. Another new setup has been developed to use a fast gas sampling probe as an ion current probe. Such a probe has been used in the dissertation to simultaneously measure the NO content inside the combustion chamber of a heavy duty diesel engine and ion current under different engine operating conditions.

AUTOBIOGRAPHICAL STATEMENT

I was born in Cairo, Egypt. I finished my school education from a well-known French school in Cairo. I completed my Bachelor of Science in Mechanical Engineering – Automotive section from Ain Shams University. Then, I finished my premaster in computer programming one year after my graduation from the college of Engineering.

I then joined EDS (Electronic Data Systems) Egypt for 3 years. I worked there as a mainframe programmer and had assignments all over the world. I traveled to Dubai, England, Belgium, Germany, Netherland, Luxemburg, and France.

In August 2005, I quit EDS and joined "Wayne State University", Detroit to pursue Ph.D in mechanical Engineering. I worked under Dr. Naeim Henein at the Center of Automotive Research as a graduate research assistant.

My area of interest is ion current sensing in diesel engines. Based on my research, I have many publications and I own 3 patents.

CPSIA information can be obtained
at www.ICGtesting.com
Printed in the USA
LVIC06n1425311013
359479LV00032B/300